P9-BJA-858

Becoming Water

Becoming Water

Glaciers in a Warming World

Michael N. Demuth

Victoria Vancouver Calgary

Rocky Mountain Books
www.rmbooks.com

Library and Archives Canada Cataloguing in Publication

Demuth, Michael, 1960-
Becoming water : glaciers in a warming world / Michael Demuth.

Includes bibliographical references.
Issued also in electronic formats.
ISBN 978-1-926855-73-8 (HTML).—ISBN 978-1-927330-47-0 (PDF)
ISBN 978-1-926855-72-1 (bound)

1. Glaciers—Canada—Popular works. 2. Glaciers--Climatic factors—Canada. 3. Global warming—Canada. 4. Climatic changes—Canada. I. Title.

GB2429.D46 2012 551.31'20971 C2011-903310-0

Printed and bound in Canada

Rocky Mountain Books acknowledges the financial support for its publishing program from the Government of Canada through the Canada Book Fund (CBF) and the province of British Columbia through the British Columbia Arts Council and the Book Publishing Tax Credit.

Canadian Heritage Patrimoine canadien

Canada Council for the Arts Conseil des Arts du Canada

The interior pages of this book have been produced on 100% post-consumer recycled paper, processed chlorine free and printed with vegetable-based dyes.

For my parents and Lydia Emily

Contents

Disclaimer

The materials presented and the views expressed or implied are not meant to reflect any view or policy of Natural Resources Canada or the Government of Canada or any of its agencies.

Introduction

I am sure I run no risk of overpraising the charm
and attractiveness of a well-fed trout stream,
every drop of water in it as bright and pure
as if the nymphs had brought it all the way from its source
in crystal goblets, and as cool as if it had been
hatched beneath a glacier.

— John Burroughs, "Speckled Trout" (1895)

Human beings are 78 per cent water. We begin our existence in what is essentially water, start drinking it before we are born, and orchestrate our lives completely around its availability. Water is essential for life. About three litres per day sustains a human life and, according to the United Nations, about 20 times that amount allows each of us to maintain basic daily hydration, sanitation and hygiene. That each Canadian on average conspicuously uses water at a rate seven times the latter amount suggests that we consider water abundant. Through geologic time and the pulsing of global climate cycles, Canadians have by some wonderful cosmic accident inherited a great deal of freshwater: currently, within our borders, about 20 per cent of that known to exist in the world. Less than half of this amount is considered renewable, or able to be replaced by natural processes at a rate that is equal to or greater than its consumption. The remainder of freshwater in Canada is non-renewable fossil water. Fossil water is that which is sealed too far under the earth's surface, too deep in the depths of lakes too large to imagine, or in groundwater reservoirs and glaciers too remote to be meaningfully measurable.

But Canadians' relationship with water is more complex than the one we have to its quantity on and under our lands. So many of us crave a view of water – of a lake or a far-flung ocean horizon, or perhaps the topography and texture of a mountain slope as water relents to gravity – because it elevates our spirits. Many simply yearn to float in it once again, hearkening to that primeval buoyancy that we enjoyed during our time in our mothers' wombs. We view it as a magnifying glass for the soul. We are almost always attracted to it, certainly to its shores, where it represents the boundary between possibilities. Paddlers, mariners, distance swimmers and free divers all relish and fear it. We often find ourselves then *in with both feet*, or playing, like budding hydrologists at work in a melting schoolyard hockey rink where ice once continuous and smooth lies in ruins made so by the inevitable force of the sun and the warm winds of spring.

A portion of our water riches lies frozen, or at melting point, and creeps along the surface of the land as glaciers. Canada's glaciers range from the vast complexes that sprawl and interconnect the deep valleys of the Coast Mountain Ranges to small opportunistic aprons of ice clinging to high-elevation cirques in the Rocky Mountain eastern slopes. Much farther northeast, ice caps nourish the large glaciers of Ellesmere Island, which calve into the Arctic Ocean or provide drinking water for an Inuit hamlet nestled in a fjord that was sculpted by the glacier itself: the *Great Carver*.

This book is about glaciers and time, about glaciers and water, and about glaciers and change. It is about the ecological and environmental importance of glaciers in the context of climate change and the future of our collective landscapes and ecosystems. I wrote it because, on the one hand, I believe the stories that our glaciers have to tell have been oversimplified by the media's preoccupation with the sound bite and propensity to dumb it down. On the other hand, formal textbooks on glaciers require a significant investment of time from readers. And so my goal was to find some middle ground from which to share the current knowledge of the stories of glaciers.

My hope is that these stories will elicit a new lens through which you might view the glaciated landscapes you walk or ski through, or those you view from an airplane or from space with Google Earth. Perhaps you might imagine yourself on a windswept mountaintop where you feel the excitement that overcomes a field glaciologist when the scene of an entire glacier reveals itself, or feel the conviction of a scientist who knows what she believes not because of any faith she might harbour but because of what she sees with her eyes.

Those of us fortunate enough to see the lands where glaciers are found – let alone live, work and play there – have our own unique feelings about them, as well as memories and names that conjure attachment, a sense of place. My goal is to place you within the texture and shape of Canada and take you on a tour of our glaciers. I often use the language of glaciology, geology and physics – words that geologist C.J. Yorath describes in his wonderful book *Where Terranes Collide* as having "contraction and metaphor." Aided by this language, we will explore the landscapes and climate that support glacier formation and the complex expressions of glaciers on the landscape. We will gain an understanding of the ways in which glaciers are nourished, flow and become part of the water that runs to the sea.

I describe how glaciers have been and are currently measured in Canada, for both domestic information needs and the study of global change. I discuss what the body of glacier data tells us in terms of cycles, trends and noise, as we contend with the driving forces of global change and adaptation. Eventually, I ask you to think like a scientist and to examine evidence of change so you may either sympathize or disagree with the conclusions you may find surrounding you.

Now that you are armed with a more thorough understanding of glaciers, I hope you will be emboldened to explore the land upstream from yourself to gain a sense of origins, and our own vulnerability. If we listen to the stories told by glaciers now, they provide hints of those they will tell to our yet unborn. They speak to us about our individual and collective relationship with becoming water.

What Is a Glacier?

We could peer into the blue depths of crevasses,
so beautiful that one might long for such a grave ...
— *Sir Leslie Stephen, "Sunset on Mont Blanc" (1873)*

Thomas Wharton, in his novel *Icefields*, details the narrative of an early explorer's relationship with glaciers. Wharton imagines glaciers as metaphors of an individual's chaotic history, of human fragility in clinging to islands of possibility, exposed by chance. Parallels link our lives to the inexorable change experienced by glaciers and, as in this manifesto, their becoming water.

Professor Graham Cogley of Trent University was more technically oriented in an article he published as part of a global effort to provide a snapshot of Earth's glaciers called *Global Land Ice Measurements from Space*. He posited that a glacier is:

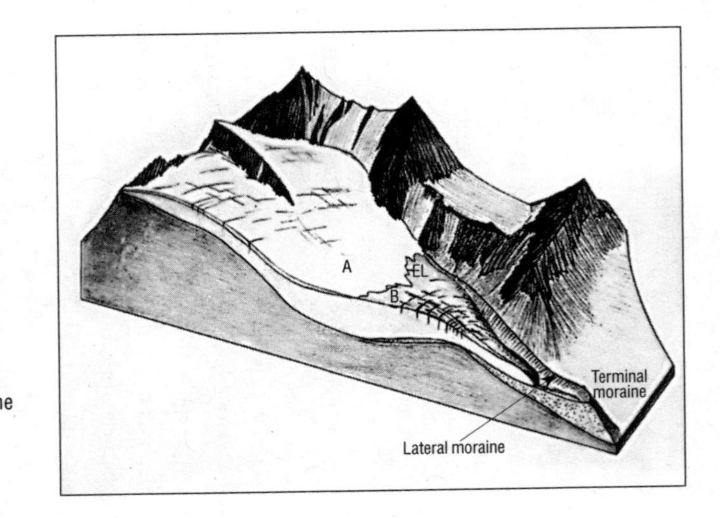

FIGURE 1: Anatomy of a typical mountain glacier, illustrating the accumulation area (A), the ablation area (B) and the equilibrium line (EL). The former extent of the glacier can be approximated by observing the position of the lateral and terminal moraines. (Adapted from the World Glacier Monitoring Service, with permission)

> … a collection of contiguous complete flowlines through snow and ice which persists on the Earth's surface for more than one year. A flowline is a sequence of ice columns of infinitesimal cross-section arranged so that each column gains mass by flow from an up-ice neighbour and loses mass to a down-ice neighbour.

Carrying forward from Professor Cogley's elegant description, we will examine the anatomy of a glacier, in particular the make-up of its surface zones. First though, let's consider the movement of glaciers.

How Glaciers Move

> I lean back on the sun-warmed rock, close my eyes, and listen. The glacier moves forward at a rate of less than one inch every hour. If I could train myself to listen at the same rate, one sound every hour, I would hear the glacier wash up against this rock island, crash like waves, and become water.
>
> — *Thomas Wharton,* Icefields *(1995)*

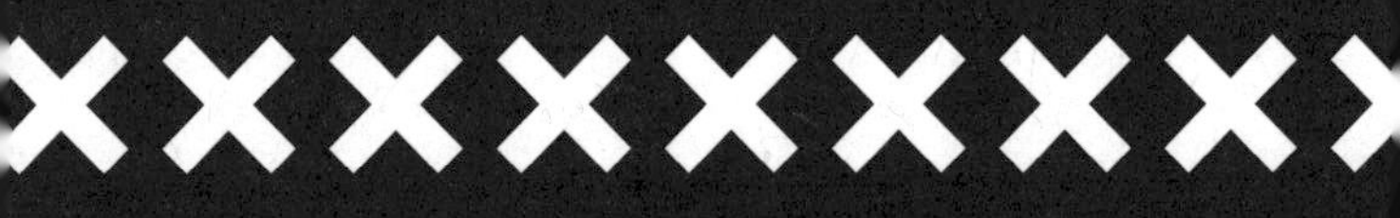

Glacier Terms and Factoids

Pronunciation – UK: GLASS-ee-ər; US: GLAY-shər

Etymology – Franco-Provençal *glace*, "ice"; from Latin *glacies*, "ice"; compare German *gletscher*, from 16c. Swiss dialect = French *glacier* or, earlier, *glacière*.

Terminology

glacier – a large, persistent body of ice originating on land – generally flowing and/or sliding due to stresses induced by its weight. Over time scales of many years, decades or centuries, a glacier forms in a location where the accumulation of snow and other forms of solid precipitation exceeds what is lost through ablation by melting, sublimation and calving.

glacial – processes and features caused by and related to glaciers.

glaciation – the process of glacier occurrence, growth, flow and form.

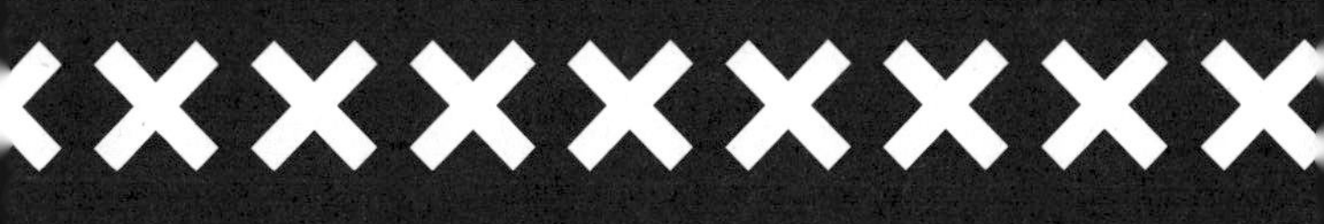

glaciated – influenced at one time by the presence of glaciers; a temporal context.

glacierized – where glaciers are currently located; a spatial context.

glaciology – traditional definition: the study of frozen water; now we define this term as the study of glaciers.

glaciologist – a person who studies glaciers.

Factoids

The ice sheets of the polar regions contain 99 per cent of Earth's glacial ice.

Glaciers are found in the mountain ranges of every continent except Australia, as well as in the high-elevation regions of the tropics.

Glacial ice is the largest reservoir of freshwater on Earth.

With the exception of high-elevation tropical settings, Canada's glacier-climate diversity exhibits the variation expressed worldwide, from the cold, arid Arctic to the warm, humid Maritimes.

After seeing a glacier, in person or through other peoples' lenses online, consider for a moment that the enormous mass of ice is actually moving. While the rates at which they move vary significantly from glacier to glacier, their movements consist of both *sliding* and *flowing*. Sliding occurs at the base of the glacier when the ice slides over the bed below. Multiple factors play roles in determining the velocity or speed of the sliding movement, including the slope and size of the glacier and the external temperature, which partially determines the amount of meltwater available to act as a lubricant for the sliding process. Flowing, or perhaps it is easier to imagine it as creeping, occurs whether or not glaciers slide. Made up of scores of individual but interconnected grains of ice that transmit the weight of the glacier to the bed, glaciers flow downhill because they continuously deform under the stress of their own weight due to gravity.

To dig a little deeper into the phenomenon of glaciers flowing, we have to briefly suspend our belief that ice is a hard, cold, brittle material – compared to, say, honey, which we view as a soft, amorphous material that flows readily. My esteemed colleague from Canada's National Research Council Nirmal Sinha, who studied the deformation of glass and ceramics at high temperatures before turning his attention to ice, actually describes ice as *hot*. He notes something called the *homologous* temperature – that is, the temperature of a solid relative to its melting point. Consider that ice, for most practical considerations, is only about 40°C (or less) from its melting temperature. Further, then, imagine what common metallic solids like steel or aluminum might be like when they are only 40°C from their melting points. At these temperatures – around 1330°C for steel and about half that for aluminum – solids become readily deformable and creep or flow under their own weight. The same holds true for ice.

Anatomy of a Glacier

The wintry clouds drop spangles on the mountains. If the thing occurred once in a century historians would chronicle and poets would sing of the event; but

Nature, prodigal of beauty, rains down her hexagonal ice-stars year by year, forming layers yards in thickness. The summer sun thaws and partially consolidates the mass. Each winter's fall is covered by that of the ensuing one, and thus the snow layer of each year has to sustain an annually augmented weight. It is more and more compacted by the pressure, and ends by being converted into the ice of a true glacier, which stretches its frozen tongue far down beyond the limits of perpetual snow. The glaciers move, and through valleys they move like rivers.

— *John Tyndall,* The Glaciers of the Alps & Mountaineering in 1861 *(1911)*

•

Generally speaking, a glacier has an *accumulation area*, over which snow mass collects throughout the year, and an *ablation* area*, where snow and ice mass is lost, that is, becomes water. Overall, if the mass gained exceeds that lost, the glacier experiences a net mass gain, or *positive mass*

* Ablation: the wasting or erosion of a glacier, iceberg, or rock by melting or the action of water, from the Latin *ablat-* (*ab-*, "off, away, from"; *lat-*, past participle stem of *ferre,* "carry")

balance. Conversely, a glacier that loses more mass than it gains experiences a net mass loss, or *negative mass balance*. Uninterrupted years of either net mass gain or loss lead to a glacier expanding or contracting its margins respectively: advancing or retreating. Glaciers don't actually retreat, they only appear to do so when more ice is melted back than is sliding and flowing down the valley.

Now imagine the accumulation of snow higher on the glacier becoming buried and eventually compressing into ice. This ice, under its own weight, flows down the valley, transported there as if on a conveyor belt, where the processes of ablation eventually free it from the glacier – as vapour, liquid or broken shards of solid ice. A glacier that is in *equilibrium* is replenishing itself by making ice in the accumulation area in equal quantities to that which is released as water or calved* away from the main part of the glacier to form spectacular ice avalanches or

* Calving: the breaking away of a mass of ice from an iceberg or glacier. Calving represents the major form of ablation from a glacial system.

simply topple into the lake or sea at the front of the glacier to become icebergs.

The process of altering the thickness, length and area of the glacier, called mass redistribution, takes time – years, decades, even centuries. The time over which a glacier will come to a new equilibrium, given a shift in the climate, will depend on the length, steepness and thickness of the glacier, as well as its thermal characteristics. If the glacier is *temperate* (at melting point), it is referred to as *warm*; if non-temperate (below melting point), it is *cold*. Generally speaking, small, warm glaciers respond more quickly to climate fluctuations than large, cold ones do.

While the processes of accumulation, flow, sliding and ablation are difficult to watch over the time frame of casual human observation, the so-called *equilibrium line*, demarcating the boundary where accumulation equals ablation, is easily recognized. For a typical mountain glacier, the equilibrium line can be identified approximately where something called the *transient snowline* is positioned at the end of summer as melting ceases and winter is dawning. The transient (impermanent) snowline advances up the glacier as snowmelt happens, exposing the darker ice of the ablation area. The rate at which this snowline rises and the ultimate position it reaches on the glacier both depend on the depth of the previous winter's snow cover and the weather conditions the glacier experiences during the summer. If winter snows are plentiful and the summer weather is cool and cloudy and perhaps even allows a heavy midsummer snowfall, the transient snowline is slow to rise and will generally be much lower on the glacier than if winter snows were marginal and the summer warm and sunny.

The position and elevation of the equilibrium line, which we now know is often demarcated by the location of the transient snowline, is an important measure of glacier health. The higher the equilibrium line is on the glacier, the smaller then is the area that is gaining mass (or nourishing the glacier) relative to the total area of the glacier, and the greater the area that is losing mass. In situations where the equilibrium line is so high that it cannot be discerned on the surface, the entire surface of the glacier is experiencing a net mass loss. Longer periods of very high equilibrium lines can herald the death knell for a glacier, because with each passing year its surface grows increasingly darker and smaller: characteristics that subsequently accelerate mass loss and glacier retreat. Called the process of *positive feedback** – somewhat deceptively, given the outcomes are not "positive" – the situation is forcing the glacier to create the conditions that speed up its ultimate destruction. Such circumstances have been recently documented for the glaciers on the eastern slopes of Canada's Rocky Mountains, where the Bow, North Saskatchewan and Athabasca Rivers are born.

While it is well beyond the scope of this manifesto to describe the physics of glacier phenomena intimately, we will entertain one basic principle that is fundamental to describing a glacier's state of mass equilibrium/disequilibrium:

* Positive Feedback: the mechanism in a system by which a process intensifies or accelerates as each cycle of operation establishes conditions that favour repetition.

the principle of mass continuity. Recall that a glacier in mass equilibrium is replenishing itself by making ice in equal quantities to that which is being released as water or through calving. Think also of how the glacier's mass is moving (flowing and sliding) down the valley or toward the ocean as though it is on a conveyor belt. If you could take a glacier, slice it at two nearby places – like cutting a slice out of the middle of a baguette – and set it aside, you would see a cross-section of that glacier in the short segment you removed. Comparing the upper and lower boundaries, or each end, of the section you cut out will reveal the state of the glacier. If the quantity of mass that flowed and slid along the "conveyor belt" across the upper boundary of the section is equal to that crossing the lower boundary, then the glacier is at a state of mass equilibrium. If more mass is leaving through the lower boundary than arrived at the upper, the glacier is growing. Following logically, if less mass is leaving that section than had arrived at the upper boundary, the glacier is shrinking.

Armed with a basic knowledge of the movement, anatomy and functioning of a glacier, let's now look at the cycle that glaciers experience. Glaciers are nourished, experience loss and subsequently provide nourishment for their surrounding environment. They are a fundamental part of our collective ecology, part of the flow of our daily lives.

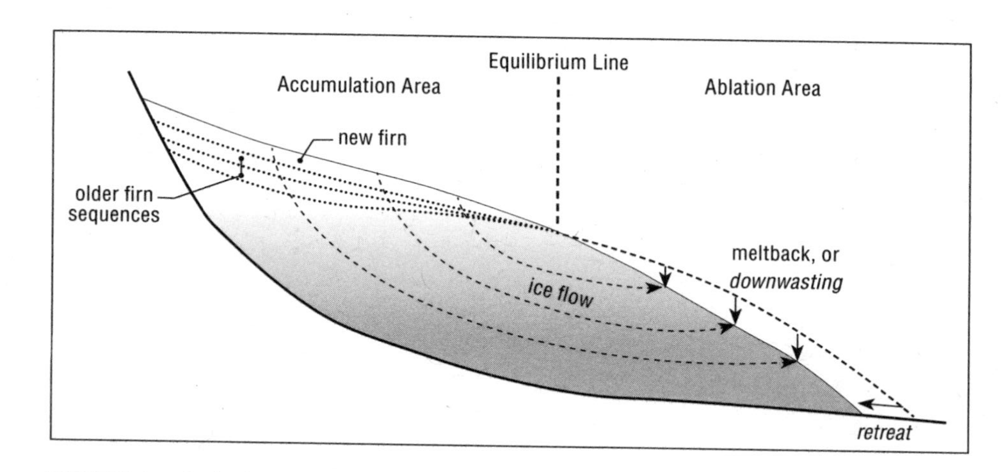

FIGURE 2: Longitudinal cross-section of a glacier, showing the approximate trajectory of a particle moving from the accumulation area to the ablation area. Snow that has survived one season is called *firn.*

The Glacier Cycle

Nourishment

Ascending up onto an icefield one eventually
comes upon a horizon of snow and sky
suspended in space between the materials
that make the nearly parallel valley walls –
themselves and this river of ice vanishing behind us
over the throw of the Earth.

At this point, we have the privilege of ascending glaciers to their upper reaches, where they are nourished. Glaciers begin high up in accumulation areas, where falling snow and other forms of solid precipitation – snow from avalanches and drifting snow – accumulate and densify into ice. The distribution of snow in mountains and over a glacier is affected by processes operating at many temporal and spatial scales. In general, regions of high elevation receive more moisture due to *orographic* uplift: moist air masses are forced upward by mountainous terrain. Snow cover is not rare at low elevations, but unless the latitude at which the glacier is located is high, such snow starts and stops and does not last long. At higher elevations, on the other hand, snow cover is often continuous for many months.

Owing to the effects on surface reflectance* and something called the *latent heat of melting* – whereby the melting process removes energy from the layers of air near the surface – the presence of continuous snow cover can lower near-surface air temperatures over several days to months. The presence of permanent snow and ice tends also to extend the duration of seasonal snow cover locally. In other words, if there is a glacier in the area, the region around it will generally be cooler and the snow will stick around a little longer than in similar areas not affected by glaciers. On a smaller scale, the accumulation of snow in an area may occur because wind

* Reflectance : the measure of the proportion of light – or other radiation – that strikes a surface and is reflected off it.

or avalanches redistribute and deposit snow based on terrain variables, such as the angle of the slope and the direction it is facing. Once deposited, snow begins its transformation into ice. Generally, the rate of densification of cold, dry, low-density snow is slower than that for higher-density snow produced by and subject to warmer conditions.

Snow that survives one year, and so must be above the equilibrium line, is called *firn*.* Under increasing stress, as the firn is buried deeper and deeper, the spaces (pore spaces) that allow air to circulate between the ice grains become less and less until finally they are no longer connected. Such compacted firn has attained the so-called *pore-close-off density* or, in relation to the depth of the glacier, the *firn-ice transition*. By this point, the ice is well on its way to the lower reaches of the glacier, where it will eventually become water.

* Firn: snow on the surface of a glacier that has remained from the previous year, which may be compact but not yet turned into ice. A transitional stage between snow and ice.

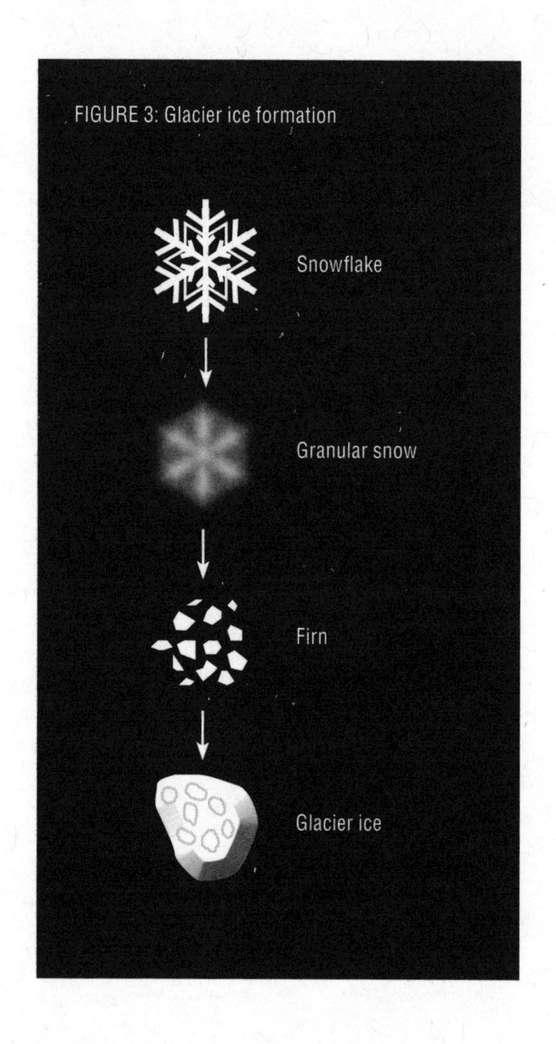
FIGURE 3: Glacier ice formation
Snowflake
Granular snow
Firn
Glacier ice

Transformation

... how you'd sigh yourself [awake]
When I raked the springtime across your sheets

— adapted from Neko Case,
"This Tornado Loves You" (2009)

As Robert Sandford articulates in his exquisite book *Water and Our Way of Life,* water is a shape shifter: "soft as a cloud and hard as a glacier; able to be transparent in all its forms." Of Earth's many thousands of different solids, none require as much heat to melt as ice does. And so a world once quickly frozen is long to thaw. The path back is unlike the one forward.

The production of liquid water on a snow-covered glacier surface begins as it does for snow everywhere. Winter is hurried off by the warmer temperatures of spring, a decidedly longer day and a warmer spectrum of light. The energy absorbed increases the snow temperature – eventually to the melting point – and also warms the surrounding air through re-radiation, which further helps melt snow cover.

Contrary to popular belief, warm, cloudy weather removes a continuous snow cover most effectively. Under clear skies, snow, having high reflectance, actually bounces most of the incoming shortwave solar radiation back into the atmosphere. Thus the sun's rays work more efficiently once the top layers of highly reflective snow begin to be removed, gradually exposing dark patches of bare ice, which absorb much more of the sun's energy (another positive feedback process). As warm air and absorbed radiation *ripens* the snowpack, water contained within it increases until it becomes saturated. At that point, the water can no longer be held in the pore spaces between the grains of snow; it must run off.

The process of meltwater production, however, varies by location on the glacier. High in the accumulation area, for instance, the bulk of the snow cover may still be well below zero when the shortwave radiation associated with the intense melting of spring begins its work. In this situation, meltwater generated at the surface percolates into the cold snow cover below and refreezes in the form of vertical *flow fingers* and horizontal ice layers, resulting in a snowpack of highly variable density. If, however, the snow cover is already warmed to the melting point, the percolation of surface meltwater occurs in a more uniform manner. Upon reaching an impeding layer, like a midwinter rain crust or the previous summer's surface, either the water will be forced downslope or its pressure will build up and force a route through the impeding layer.

Over the lower reaches of the glacier, meltwater that reaches an impeding layer or the glacier ice surface itself runs off unseen. As the snowpack is removed and the transient snowline works its way up the glacier, however, snowmelt conspires with increasing amounts of icemelt to generate runoff on top of the glacier (supraglacial), which at times looks similar to water flowing across an impenetrable tarmac during a summer downpour.

The ice on the surface of the glacier is not uniform, though, and its cracks, crevasses and other discontinuities often divert the runoff water from its desired path along the fall line.* The combination of runoff water and the physical characteristics of the glacier surface can

* Fall line: the line of steepest descent on a slope.

generate small whirlpools that give way to larger vortexes, which in turn hydraulically* drill their way into the glacier and form vertical shafts called *moulins* or mill holes.

As the melt season advances, the water within the glacier (englacial), such as that in moulins, continues to labour and subsequently produces a complex labyrinth of tunnels called *plumbing*. The plumbing created within a glacier eventually provides a route for water to reach the glacier bed (become subglacial) and find its way out to contribute – along with supraglacial and englacial runoff – to the creation of a proglacial stream: a stream immediately in front of the glacier. The proglacial stream takes the water downvalley, where it is joined by other streams to form ever-larger rivers leading to the sea. These coalescing source waters nurture natural and human systems that have adapted to, or have been designed around, the availability of flowing water.

* Hydraulic: denoting or relating to a liquid moving in a confined space under pressure.

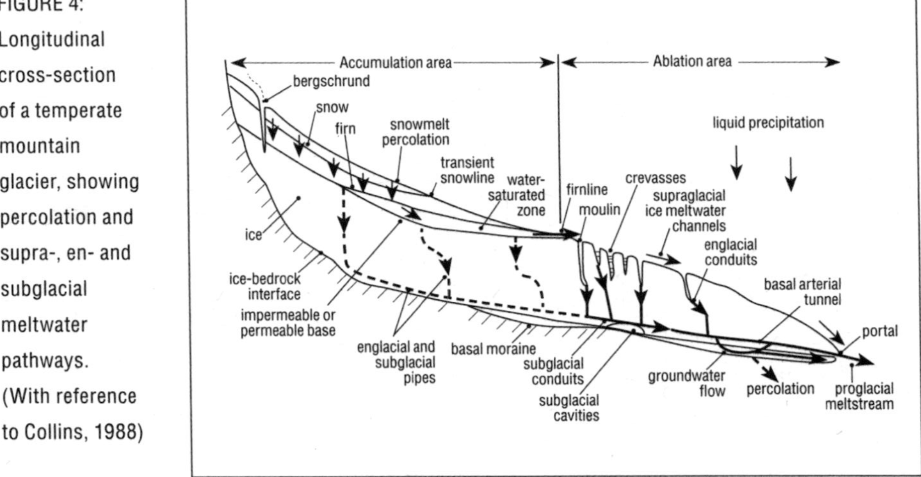

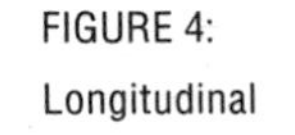

FIGURE 4: Longitudinal cross-section of a temperate mountain glacier, showing percolation and supra-, en- and subglacial meltwater pathways. (With reference to Collins, 1988)

Deh Cho: Big River, Flowing Much

A river is life and light, especially in timbered country; it is always the easiest natural way of travel and it is used as such by many creatures besides man. No clean river can be other than beautiful and it has changing beauty. Even a streamlet can become impressive in floodtime and the greatest river has a light, almost intimate quality of gentleness at its lowest Summer level.

— *Roderick Haig-Brown,* Measure of the Year *(1950)*

When under the influence of glaciers, the continuous flow of a river is preserved even during dry weather when, without glacial input, it may become gentle. On the other end of the spectrum, extreme heat and rainstorms can amplify already turbulent rivers that make their way from glaciers in the summer months. On balance, though, glaciers act as regulators of river flow, smoothing out fluctuations throughout the year.

Over the scales of climate variability, glaciers store water during cool, wet periods (generating positive mass balances) and release it during warm, dry ones (generating negative mass balances). On an annual scale, meltwater from glaciers extends the seasonal flow peak that would otherwise occur only in association with spring snowmelt. As the snowpack reservoir becomes depleted, and, for instance, if it happens to be a low precipitation year in the region, mountain rivers where glaciers are present are maintained, in part, by direct glacier meltwater contributions.

When glacial meltwater dominates the streamflow, the river becomes *turbid:* cloudy or thick with suspended matter. This suspended matter – the finely powdered rock formed by glacial erosion, called glacial *rock flour* – is often found suspended in the water of mountain lakes, giving them their turquoise appearance. As the glaciers vanish, we are compelled to wonder whether mountain lakes will look different and, if so, what changes we will witness.

Another question that is often raised ponders the idea of whether or not we should expect more water from glaciers if indeed the climate is warming. The answer is yes and no, together. Yes, in that from one summer to the next, if it is warmer in that second summer than in the first,

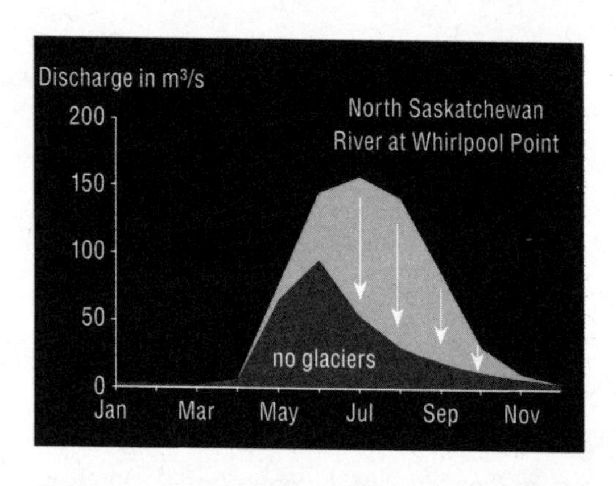

FIGURE 5: What would happen to glacier-fed rivers if their already rapidly shrinking glaciers just disappeared? A distributed hydrological model called watflood was used to construct this hydrograph of just such a situation. Note how the peak flow would occur earlier in the year, and that late-summer flows would be severely diminished. The flow, or discharge, of a watercourse is measured in cubic metres per second. This figure was adapted from a modelling study conducted for Alberta Environment that I co-led with my colleague Alain Pietroniro from Environment Canada. The figure was expertly crafted from the original data by Scott St. George, formerly with the Geological Survey of Canada, and University of Saskatchewan M.Sc. student Laura Comeau.

you will experience greater melt and *yield* from the glacier. And no, in that if over the long term the glacier is in a state of contraction – so that over, say, 50 years, it will have become significantly smaller – despite warmer temperatures now than then, the yield of the glacier will be less even when the climate is warming. This example represents the dynamics – differentiating the different scales, trends and cycles – involved in questions surrounding climate change. We will address these issues in greater detail in the chapter "Becoming Water."

Glacier-fed rivers are vital to our ecosystems, as they provide unique habitats within the river and on its edges. Ice becoming water influences the water temperature and provides cold waters that many invertebrate species and fish are highly adapted to. In addition, glacier-fed rivers are hydraulically dynamic, with the glacier continually providing sediment to the river, creating sand and gravel bars and influencing the shape of the river over time. The thick sediment deposits in the streambed lower the channel's carrying capacity, which causes the river to flood during periods of high streamflow, maintaining healthy riparian zones.*

Flow to Oceans

And it's good and it's true, let it wash over you
Untethered and without a reason

— *Great Lake Swimmers, "River's Edge" (2009)*

Mountain glaciers and Earth's great ice sheets play important roles in sea-level change through the meltwater that they release, which flows to the oceans. Of all the factors contributing to changes in sea level, including the warming and resulting thermal expansion of the oceans, meltwater from ice on the land exerts the largest influence. To understand this phenomenon with more clarity, a distinction must be made between the effects of ice sheets and those of mountain glaciers. The outcomes of ice sheets contributing to sea-level change will be witnessed over long time periods, whereas, while relatively less significant to sea-level change,

* Riparian zone: a stream and all of the vegetation on its banks.

outcomes of mountain glaciers are experienced over a shorter time frame. In fact, recent analysis has determined that the melting of large glacier systems in Alaska, Yukon and northwest British Columbia from 1995 to 2001 contributed more to sea-level rise than did the melting of the Greenland Ice Sheet over the same period. Because of the pervasive impacts of sea-level rise, particularly on the people least able to adapt, all contributions must be assessed with rigour.

Sea-level change is a complex phenomenon, with global average increases in sea level due to glacier and ice-sheet melting being somewhat misleading in the regional context due to local changes at the source of the meltwater. Two things happen locally when glaciers and ice sheets lose mass: because there is less mass, the gravity field is reduced; and with a lesser load, the land surface also rises. The net effect is that, relative to the land surface elevation, local sea level actually decreases. In fact, recent analysis shows that the increased amount of meltwater from Greenland or Antarctica is manifesting its effect on sea-level rise great distances away, while near the ice sheet sea level is actually declining.

One of the short-term concerns about warming ice sheets is that their vast, fast-flowing regions – for example the Jakobshavn ice stream on the west coast of Greenland – could, with increasing meltwater making its way to the glacier bed (and in combination with rising sea levels generally), allow enormous quantities of ice to enter the ocean and raise sea levels significantly and quickly. The consequences of glacier-climate changes make it necessary for all people, glaciologists in particular, to recognize and pay attention to change. In Canada, our glaciers are telling their own stories of change. If we listen closely, they even whisper hints of the future.

Canada's Glaciers: What Was, What Is

The glaciers made you, and now you're mine
— *Great Lake Swimmers, "Your Rocky Spine" (2007)*

The Great Carver

Canada's climate and landscape currently support some 200,000 square kilometres of glacier cover, more than any other country in the world outside of the great ice sheets of Antarctica and Greenland. About 150,000 square kilometres resides on several Canadian Arctic Islands, home to some of the largest polar icefields in the world. Another approximately 50,000 square kilometres of glacier cover is located in Canada's western and Northern *Cordillera** – specifically

* Cordillera: a system or group of parallel mountain ranges together with the intervening plateaux and other features.

in the Rocky Mountains, Interior Ranges, the Selwyn and Mackenzie Mountains of the Northwest Territories and the Yukon, and along the Pacific Coast, where many of Alaska's great glaciers are born. This latter region, spanning the border between Canada and the US (Alaska), also includes the world's largest non-polar icefield, which adjoins Canada's highest peak, Mount Logan (5,959 metres), in south-western Yukon.

Canada's icefields and glaciers derive from the Laurentide and Cordilleran Ice Sheets of the last glacial episode of the *Pleistocene epoch*.* These glaciers are a combination of remnants of old glaciers that could not re-form under present conditions if they were to disappear, and ones that are still viable on account of their location in the topography and today's regional climates. Before we explore these regional settings in some detail, let us have a brief look back at the

* Pleistocene: from the Greek *pleistos*, "most," and *kainos*, "new" or "recent," pertaining to a time of glaciation; *epoch*, of course, refers to a span of time whose character is distinct or identified from a fixed point.

last glacial episode, when virtually all of Canada was covered by ice.

The last major advance of continental glaciers that made up the Laurentide Ice Sheet is termed the *Wisconsinan Glacial Episode* (WGE).* Glacial episodes are separated by warmer interglacial episodes, like the one we currently enjoy. Glacial episodes are sometimes erroneously referred to as ice ages. An ice age actually refers to periods during which the Earth's surface features ice covers that are continental in scale. Because the great ice sheets of Antarctica and Greenland still exist, we are still in an ice age – the Quaternary Ice Age. This ice age began 2.588 million years ago, at the start of the Pleistocene epoch, during which most of Earth's recent, repeated glaciations took place. Glacials are episodes of extra-cold climate within an ice age

* Why Wisconsinan? We name glacial episodes in association with the contemporary location of regional evidence for them on each continent. The Wisconsinan equivalent in northern Europe, for example, is called the Weichselian, corresponding to the Weichsel (Vistula) River.

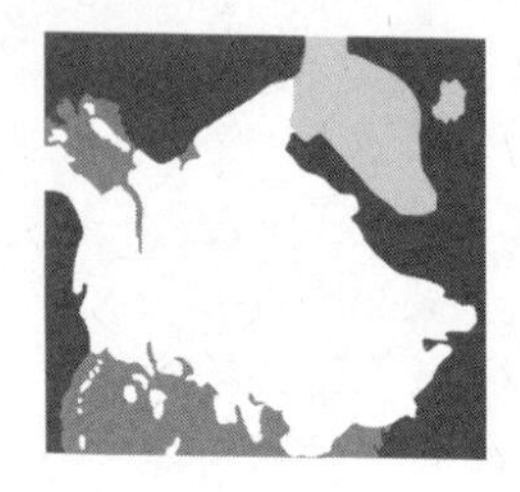

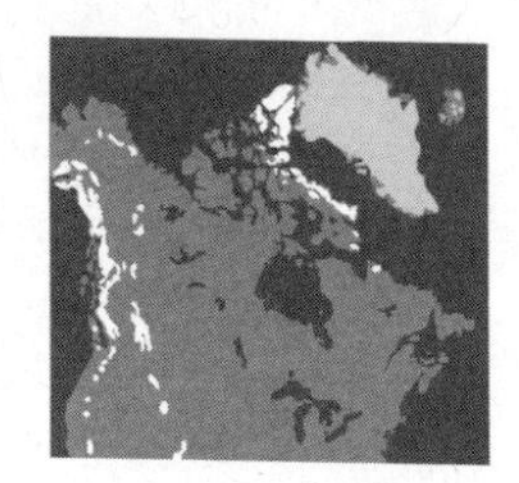

FIGURE 6: The extents of the ice sheets of Cordillera and Laurentia at approximately 12,000 years before present.* Also shown are the more isolated but significant regions of late Pleistocene glaciation in the northwestern US, and that inundating what is now the Brooks Range in Alaska.

FIGURE 7: The approximate locations of present-day glacier cover in the North American Cordillera and the Canadian Arctic Islands.

* A.S. Dyke, "Late Quaternary Vegetation History of Northern North America Based on Pollen, Macrofossil, and Faunal Remains." Animated maps of the post-Wisconsinan Glacial Episode ice sheet retreat are available at http://is.gd/nOIjGe and http://is.gd/YVouy4. See also the "Bookshelf" entry "Retreat of the Laurentide Ice Sheet."

during which glacier cover is very extensive on most continents in both hemispheres. Glacials themselves have periods of more or less cold corresponding to glacier limits that are more or less extensive – the warmer interval referred to as an inter-stadial.

The WGE began about 110,000 years ago, reaching its greatest advance 21,000 years ago and ending about the same time as the Pleistocene epoch, around 12,000 years ago. The advance was encouraged by a period of cooling climate, in which the glaciers of the mountain West grew and coalesced to form broad mountain ice sheets that eventually conspired to form the Cordilleran Ice Sheet, covering most of British Columbia and southern Yukon and portions of Alaska, Montana, Washington and Alberta. As they flowed and carved downward and outward, they created great valleys, eventually emerging on the Canadian Plains to meet the Laurentide Ice Sheet, which, in part, was plowing and spreading westward from various centres of initiation in the Hudson Bay region. Between the Cordilleran and Laurentide Ice Sheets, virtually all of Canada was covered by ice.

The WGE altered the geography of North America north of the Ohio River in a big way. At its maximum stage, it also reached down to the US Upper Midwest and New England. In Ontario's Killarney region and other parts of Georgian Bay and Lake Huron, you can easily observe the grooves etched in the bedrock left by these glaciers. In southwestern Saskatchewan and southeastern Alberta, the coming together of the ice sheets of Cordillera and Laurentia formed the Cypress Hills, located at the northernmost point in the continent that remained south of these ice sheets. Their southerly manifestations include *terminal moraines** that form Long Island, Cape Cod, Martha's Vineyard and Nantucket in the US, as well as the Oak Ridges Moraine in south-central Ontario. Of course, in the regional namesake itself, Wisconsin, the retreat of the last Wisconsinan ice left the well-studied Kettle Moraine.

* Terminal moraines: cross-valley/flow-line-oriented ridges of material formed by the pushing action of the glacier or ice sheet margins.

As the Cordilleran Ice Sheet withdrew, it shaped the great gorge now called the Okanagan Valley in BC. And when the Laurentide Ice Sheet began to retreat, it scoured the land and pooled meltwater at its margins to form the Great Lakes. Imagine that the Laurentide Ice Sheet was almost four kilometres thick. The pressure of its weight on the land was immense; even today, the land in the Great Lakes Basin continues to rebound and rise over seven centimetres every hundred years.

Beyond dramatically changing the geography of Canada, glaciers and the movement of ice sheets also served to pick up and move materials. Consolidated material is made up of substances that cannot be easily picked up and transported by the glacier, whereas unconsolidated material, like soil and gravels, is readily able to catch a ride with the glaciers. Moving unconsolidated materials from one space to another, glaciers left immense deposits of redistributed drift minerals, for instance, in their wakes. You will recall that vast but poorly quantified groundwater resources can be contained within these glacial deposits.

At the end of the WGE, Earth entered into the *Holocene,** the epoch that continues today. The early Holocene was largely characterized by an episode of exceptionally warm climate (between about 5,000 and 9,000 years ago) that is referred to as the *Holocene Thermal Maximum*, the *Climatic Optimum* or the *Hypsithermal*. Under this protracted warming, these great glaciers and ice sheets melted and withdrew, sea level rose globally and the previously inundated land, now free of its heavy covering of ice, continued to rebound.

The climate conditions of the early Holocene allowed for the dawn and expansion of modern civilizations. The *Medieval Warm Period* (950–1250 CE) and the so-called "Little Ice Age" or *Neoglacial* (1500s–1800s CE, with regional variations) are two additional and significant climatic expressions that influenced and challenged the adaptive capacity of societies and other natural systems. More recently, we have begun using the term *Anthropocene* to describe an adjunct period

* Holocene: from the Greek *holo*, "whole," and *kainos*, "new" or "recent."

in the Holocene during which human beings are exerting an influence on the Earth's stratigraphic* record as borne out by our activities and their effect on the land surface, climate and ecological processes. For example, in the record recovered from a glacial ice core, we can detect lead used as an anti-knock agent in gasoline, and even see its decline during the period when this additive was phased out of gasoline production. Variations in atmospheric carbon dioxide levels can also be recovered from ice cores and other deposited stratigraphic sequences.

What we know about global average temperatures during the Holocene is largely derived from what is called *proxy*** evidence – in this case, indirect estimates of temperature taken from signatures (biophysical or chemical) left in glaciers, trees, coral reefs and lake and ocean sediments – as well as values taken from direct

* Stratigraphic: relating to or determined by the layers (strata) within a deposition such as sediment, sedimentary rock, a glacier or a snow cover.

** Proxy: a figure that can be used to represent the value of something in a calculation.

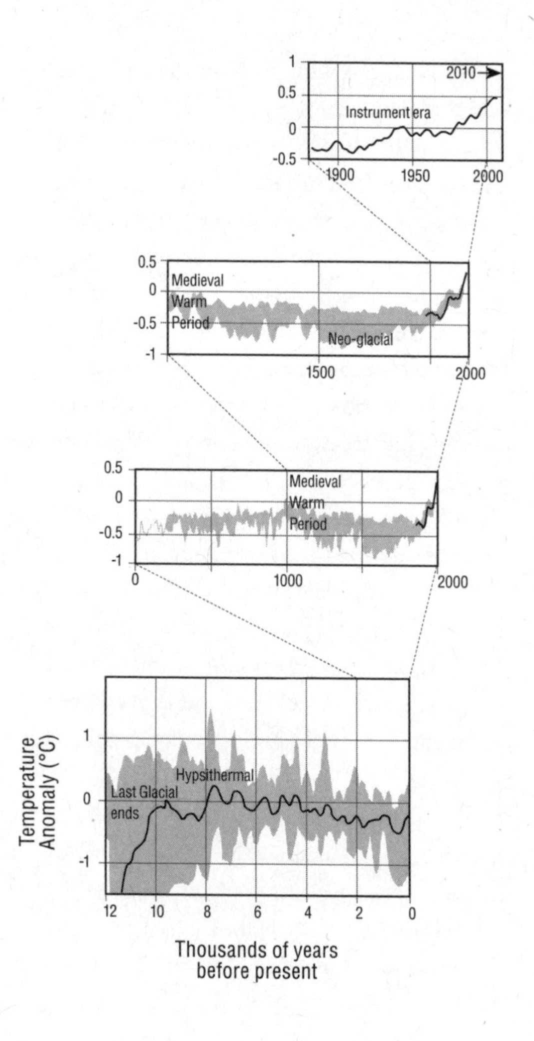
1
0.5
0
-0.5
2010
Instrument era
1900
1950
2000
0.5
0
-0.5
-1
Medieval
Warm
Period
Neo-glacial
1500
2000
0.5
0
-0.5
-1
Medieval
Warm
Period
0
1000
2000
Temperature Anomaly (°C)
1
0
-1
Hypsithermal
Last Glacial
ends
12
10
8
6
4
2
0
Thousands of years
before present

FIGURE 8: The evolution of Holocene global temperature as reconstructed from indirect proxy evidence and from direct instrument measurements. Shaded regions illustrate the range of the available reconstructions. The solid line in the lower panel represents the average of the whole Holocene proxy record – eight records, broadly describing the main characteristics of Holocene variability, each having first been smoothed to produce a resolution of 300 years. The temperature anomaly shown is relative to the average temperature of the mid-20th century. The range of the reconstructions for the most recent 2,000 and 1,000 years (middle panels) is based on 10 higher-fidelity records, each smoothed to a resolution of 10 years. Here the temperature anomaly is relative to the average for the instrumental era, beginning in 1880 (represented by the solid lines, smoothed to reveal five-year variability). The reference arrow describes the global value for 2010. (Adapted from and using associated references and data sources from NASA Goddard Institute for Space Studies and http://en.wikipedia.org/wiki/Temperature_record)

instrument-based measurements that began around 1880 CE. We will discuss records of changing climate later, but at this juncture it is worth noting – while acknowledging the fact that there may have been regional or global fluctuations in temperature that are unrepresented by indirect proxy evidence – that global average temperatures appear to now be achieving values outside the bounds of Holocene variability. This rise in global temperatures has taken place during the time in which modern civilizations have been established and have adapted.

Now that we have at least some sense of glacial space and time, our next journey is over the vast frontier regions where the glaciers of Canada presently lie.

The Arctic Islands

I have waited with a glacier's patience

— *Neko Case, "This Tornado Loves You" (2009)*

Qausuittuq (Inuktitut): *The Place with No Dawn*

The Queen Elizabeth Islands in Canada's High Arctic is an archipelago made up of the major islands of Ellesmere, Meighen, Axel Heiberg,

Canada's Arctic Islands

Island	Area	Rank	Population
Baffin	507,451 km^2 (195,928 sq mi)	Largest island in Canada and fifth-largest in the world	10,745 (2006)
Ellesmere	196,235 km^2 (75,767 sq mi)	World's 10th-largest island and Canada's third-largest island	146 (2006)
Devon	55,247 km^2 (21,331 sq mi)	Canada's sixth-largest island, and the 27th-largest island in the world	uninhabited
Axel Heiberg	43,178 km^2 (16,671 sq mi)	31st-largest island in the world and Canada's seventh-largest island	uninhabited
Bylot	11,067 km^2 (4,273 sq mi)	71st-largest island in the world and Canada's 17th-largest island	uninhabited
Meighen	955 km^2 (369 sq mi)		uninhabited
North Kent	590 km^2 (230 sq mi)		uninhabited
Coburg	411 km^2 (158.7 sq mi)		uninhabited

Coburg, Devon and North Kent. They are all in Nunavut, with Melville lying partly in the Northwest Territories. It is on these islands that Canada's High Arctic glaciers reside. The largest, Ellesmere Island, is an expanse of cordilleras where immense icefields partially inundate vast subglacial mountain ranges and flow outward to nourish outlet glaciers in much the same way that the iconic Athabasca Glacier in Alberta drains a portion of the ice generated by the Columbia Icefield. Many of Ellesmere's outlet glaciers end abruptly at the sea, shedding ice along impressive calving fronts.

Many glaciers in the Queen Elizabeth Islands are classified as *ice caps*, a term often used to describe the Arctic Ocean sea ice. In glaciology, an ice cap specifically refers to a dome-shaped ice mass that inundates the landscape with ice flowing radially out from a distinct summit to the lower reaches of the landscape. An archetypal ice cap is the well-studied Devon Ice Cap.

There are also many smaller ice masses that belong to a family of glaciers that could not re-form in today's climate if they were to disappear. Some of these behave as *glacier reservoirs*, a term

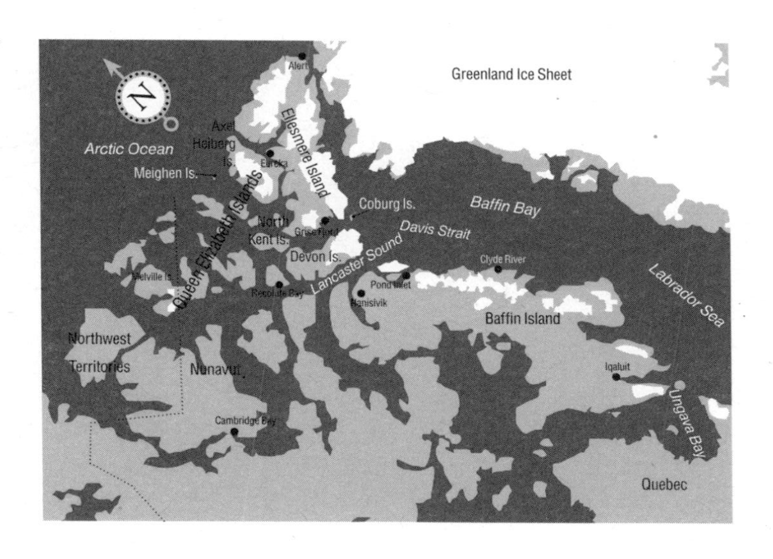

FIGURE 9: The glacier cover of Canada's Arctic Islands and northern Labrador. Note the proximity of Ellesmere Island's great ice caps to Greenland.

coined by celebrated French glaciologist Louis Lliboutry. Glacier reservoirs have a geometry and placement on the landscape such that they do not flow from higher to lower elevations but rather are termed *stagnant*, with their volumes fluctuating only on account of losses or gains in the elevation of their surface – like a hydro power storage reservoir's level would fall or rise. One of these is Meighen Ice Cap on Meighen Island. Such glaciers are considered ideal for the study of climate because they are free from the complications of glacier dynamics that redistribute ice by flow and sliding.

Exempt from these complexities of glacier dynamics though it is, however, Meighen Ice Cap does respond to the dynamics of the surrounding ocean, as do other glaciers whose margins are close to the sea. The presence or absence of summertime sea ice is particularly important. Whereas sea ice near the land was once a reliable feature of the summer months, it is increasingly absent in the High Arctic, leaving stretches of open water to freely influence weather in the area. Drifting sea fog, for instance, is common over open water and can reduce melting over the lower reaches of nearby glaciers during a season when typically higher rates of melting should occur.

Summer and autumn conditions control the general behaviour of High Arctic glaciers. It is in these seasons that much of the snow falls at higher elevations, with relatively little in the very cold, dry days of winter. Glaciers in the High Arctic are generally frozen to the bed, so rather than sliding along the surface they deform and flow. Recent glacial analysis conducted by the Geological Survey of Canada reveals that over the last quarter century this region has experienced some of the warmest weather of the past 4,000 years. Yet because its glaciers are largely "cold" (non-temperate or at temperatures below melting point), only very small changes in the area and volume of these glaciers have been documented. They respond ever so slowly. In fact, the extra cold of the Little Ice Age (1500s–1800s CE) may just now be being felt – by conduction – at the beds of the upper regions of these glaciers.

But very small changes over very large areas imply significant impacts over time. In the difficult-to-study, inhospitable High Arctic region of Canada, the subject of much glaciological research and monitoring is discerning patterns and trends in these small changes using increasingly more accurate instruments and technologies. In a nutshell, High Arctic glaciers are subject to both long-past and contemporary conditions simultaneously. Because of this extraordinary fact, these glaciers need to be subject to mindful observations and analyses so that we can hear the stories that they are telling.

The northern shores of Ellesmere Island are also home to *ice shelves*. Ice shelves are massive floating sheets of ice that are permanently attached to a land mass. They are essentially glaciers flowing out onto the sea, helped along by their interactions with the coastal climate and the salinity, or concentration of dissolved salts, of the waters, which helps them float. If an entire ice shelf or even a portion of it breaks free from land, it becomes an *ice island* that floats amid the sea ice or drifts at the whim of wind and water currents. Breaking free is a complex process. Imagine the ice shelf rising up and down with the tide. In doing so, it forms a *hinge zone*, which is similar to the fold line that develops if you flex a piece of plastic back and forth. Eventually a series of cracks can be seen along the hinge zone, though these by themselves may not result in the ice shelf breaking free. If there is an absence of sea ice (which would otherwise keep the ice shelf in place) and a persistent offshore wind, the cracks could combine with those factors to help the ice shelf drift offshore and become an ice island.

Ice islands have garnered attention recently, such as in mid-2011, when a massive ice island was drifting off the coast of Labrador and dazzling researchers and spectators with its mountains, valleys, waterfalls and even seals. With peaks 30 metres high, the ice island was over five kilometres long – nearly 50 square kilometres large in total – by the time it edged toward Canada's coast. A year earlier, when the ice island broke off of Greenland's Petermann Glacier, it was over five times that size. Given the truly awe-inspiring visuals, as well as the potential causes and consequences for such ice islands, they tend to conjure images of cataclysmic events, disintegration and sometimes opportunity. To the later point, some of these islands have even been used as magnificent floating laboratories, which, as they drift, allow detailed studies of our great polar ocean.

Auyuittuq (Inuktitut): *The Land that Never Melts*

To the south of the Queen Elizabeth Islands lie Baffin and Bylot Islands in Canada's Low Arctic. Bylot Island, lying northeast of Baffin Island and separated from it by a channel called Navy Board Inlet, exhibits a labyrinth of large glacier systems covering almost half of its area. On Baffin Island, the distribution of the glaciers is very different than the nearly ubiquitous glacier cover of the Queen Elizabeth Islands. Here, glaciers are found primarily amongst the cordilleras of Baffin's northern and eastern coastal regions. Icefields nourish a vast array of outlet glaciers, many of which reach down to Davis Strait through spectacular fjord-like channels.

The interior of Baffin Island is devoid of glacier cover, with the exception of several small stagnant ice caps and patches of ice in the north, and two large ice caps called Barnes and Penny, respectively, in central and southern Baffin Island. The Barnes and Penny Ice Caps exhibit ice on their margins that has been determined to originate from the Laurentide Ice Sheet. Barnes Ice Cap is an isolated ice mass splayed out in a simple configuration over an upland area in the central region of the island, whereas Penny Ice Cap, well south and on the Cumberland Peninsula, might be considered a hybrid, exhibiting both simple configurations along some of its margins and a vast icefield feeding numerous outlet glaciers elsewhere.

Torngat (Inuktitut): *Place of Spirits*

Leaving the Arctic Archipelago and examining the northern reaches of Newfoundland-Labrador, you will find the southernmost glaciers in eastern North America. The highest peaks of the Torngat Mountains hide groups of glaciers that cling to, and are protected by, cirque valley walls. In this "place of spirits," it was likely that Maritime Archaic Indians, Pre-Dorset and Dorset Paleo-Eskimos and the Thule culture, which merged into today's Inuit culture, witnessed the fluctuation of these glaciers over a time span of at least 7,000 years. The landscape here is what we would consider textbook glaciation, with U-shaped valleys and archetypal pyramid-shaped peaks, along with many other features that are typically formed by glacial erosion. The region's glaciers, now very small and in many cases covered with rock debris fallen from the surrounding steep mountain walls, dependably nourish the groundwater-laden land in front of glaciers. These glaciers collaborate with abundant precipitation to interact with frozen ground and influence a marvellous system of pristine lakes and streams. Though uniquely breathtaking, the Torngats are a rarely visited Canadian national park. Beholding this landscape continues to be the privilege of very few people.

The Mountain West

For the Land having been
to the Ocean's depths and
reaching out Her arms for the Sky

We cast our gaze now toward the much younger mountain West, where current plate tectonics theory holds that, some 200–300 million years ago, a series of massive Pacific islands were

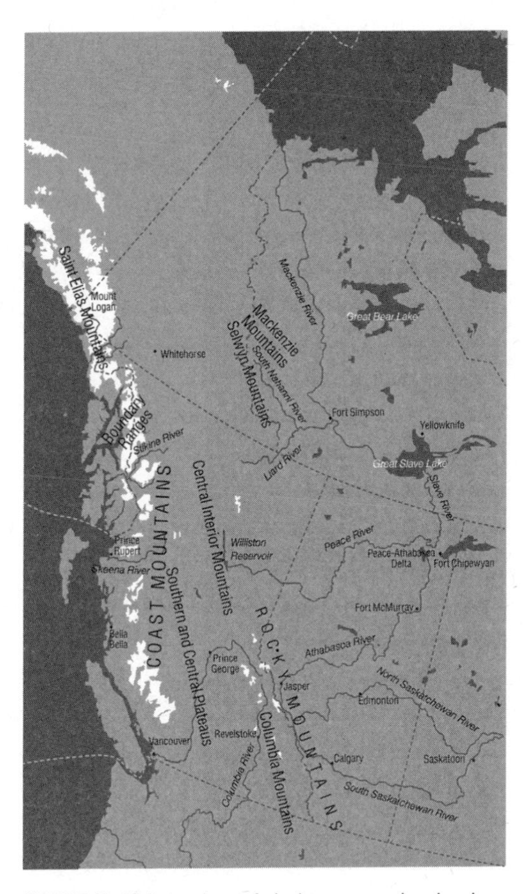

FIGURE 10: Major regions of glacier cover and major river systems and settlements of Canada's mountain West.

carried on the Pacific Plate in a northeasterly direction, colliding with the North American Plate, which was moving west. This formed a series of belts* that roughly correspond to the textures and facets expressed over the mountain West's Coastal, Western, Interior and Eastern regions. While the Arctic Islands, in particular the Queen Elizabeth Islands, are heavily ice encrusted and could eventually reveal vast groups of mountains, the West's relatively younger collisions and more rapid deglaciation has revealed huge diversity over short distances. Given such complexity, efforts to generalize the natural environment are thwarted. Instead of the full picture, we will do what we can to look at snapshots in order to increase our understanding of the region.

Deisleen (Tlingit): *Long, Narrow Water*

The Coast Mountains extend from southwestern Yukon through the Alaska Panhandle and

*Orogenic belts are ranges of mountains that are created by folding and other deformation where two continental plates collide and push upwards.

down the coast of British Columbia. Subject to moisture-laden air masses driving up steep mountain gradients – in some cases 5,000 metres of relief only 12 kilometres from the sea – these mountains are the dumping grounds for vast quantities of snow. The northern Coast Mountains include the venerable Saint Elias Mountains, whose Himalayan-scale peaks rise out of vast oceans of ice, including several of the highest peaks on the continent: the second-highest being Mount Logan (5,959 metres); the fourth, Mount Saint Elias (5,489 metres); and the seventh, Mount Lucania (5,260 metres).

Along the Canada-US border, one finds countless valley glaciers and large icefields that transition into outlet glaciers; several of the large outlet glaciers that contribute water into the Pacific Ocean via Alaska are in fact born and nourished in Canada. Along the eastern margins of this highly glacierized region are several eastward-flowing outlet glaciers that intersect north–south-oriented river valleys. In several instances, the ice has advanced to block the valley and has impounded vast quantities of water. When such configurations evolve, sudden releases of water, or *jökulhlaups**, act like dams bursting and pose considerable hazards to the downstream environment.

The region also exhibits numerous *surge-type* glaciers, which experience temporary episodes of *fast flow,* or *rapid movement*, perhaps some 100 times quicker than normal. Several surge-type glaciers have played roles in creating *glacier-dammed lakes.* While generally unpredictable, surges sometimes occur on a regular basis. The Saint Elias Mountains are a renowned laboratory for the study of surge-type glaciers. They have also attracted the interest of leading scientists in a quest to unravel the region's complex geological, climate and glacial history – their lofty summits provide archives of snow and ice unaffected by melt but subject to the regional and global circulation of water vapour and contaminants. These mountains and glaciers are a training ground for Himalaya-bound mountain climbers and, like their smaller cousins in the

* *Jökulhlaups*: the Icelandic term for a glacier-burst – used worldwide to describe sudden flood-releases of meltwater from glaciers and ice sheets.

Arctic Islands, they attract alpinists from all over the world.

Working our way down the coast and along the Alaskan Panhandle, we find the Boundary Ranges. Here the glacier configurations are somewhat similar to those of the Saint Elias Mountains. Snow and steep environmental gradients act together to produce some of the most spectacular glaciers on Earth, some of which flow through coastal rainforest environments and contribute to great rivers bound for coastal fjords that reach out to the celebrated Inside Passage.

Farther south, the glaciers of the Coast Mountains thin out in the region between Prince Rupert and Bella Bella, and though still plentiful, they are less extensive. However, in the Bella Coola, Chilcotin and Homathko regions, glaciers are greater in size and grandeur.

In these areas, large icefields and the deep valleys between them provide for some of the most breathtaking maritime alpine terrain in the world – in my estimation, the twin of New Zealand's Alps and the *Ata Whenua* (Shadowland) that is her fjordland. This region is also home to the highest peak that lies entirely in BC, Mount Waddington (4,019 metres), which is surpassed in BC only by the US-Canada shared Mounts Fairweather and Quincy Adams.

Queneesh (K'ómoks): *The Great White Whale*

There are also glaciers on Vancouver Island: 40-odd, in fact, which were first studied in the 1970s in association with the control they exert on the water temperature of streams that support the hatching and rearing habitat of salmon. These glaciers lie in the very mountainous east-central region of the island, with the largest, Queneesh, entrenched in the stories of water and the landscape held sacred by the K'ómoks First Nation. Part of the legend is that during a great flood of the Comox Valley, four canoes full of K'ómoks people saw a large white whale and attached ropes to the creature so that they could not be washed away in the flood. Once the waters fell back, however, the whale was stranded in a lake that later froze to become Queneesh, the Great White Whale.

Illecillewaet (Shuswap): *Swift-Flowing Water*

As air masses move farther eastward beyond the rain shadow* that is the southern and central plateaux of the British Columbia interior – think mountain desert near Osoyoos, horticulture in Okanagan country and cattle ranching near Prince George – they are once again sent aloft by the Columbia Mountains in the south and the varied mountains of the central interior. The Columbia Mountains extend northward from the Slocan and Kootenay regions of southern British Columbia, through the Monashee, Selkirk and Caribou Mountains, all encircled by the great arch of the Big Bend of the Columbia River. The central interior mountains extend toward the Yukon, segmented by the Fraser, Skeena, Stikine, Liard and Peace River basins, within which lie the Omineca, Skeena, Hazelton and Cassiar Mountains and Stikine Plateau. A vast array of glaciers and icefields,

* Rain shadow: an area on the leeward side of mountains or other high ground, having relatively little precipitation because clouds release their moisture on the windward side.

which have only recently been inventoried in detail, reside here. While alpine ski-touring, climbing and helicopter-accessible skiing are popular here, these regions, broadly speaking, are some of the wildest in the mountain West because they are transected in only a few places by major east–west railways and highways.

Bounding the region to the east is the Rocky Mountain Trench, a large fault-induced valley superimposed by classic glacier-carved topography, extending approximately 1,600 kilimetres from Montana to just south of the British Columbia–Yukon boundary. So pervasive a feature is it that aviators dead reckon to it on their northward or southward transits, and it can be seen with the naked eye from the moon. That said, you might miss its start, middle and end on your traverse across its 25-kilometre width during your flight from Calgary to Vancouver!

As-sin-wati (Cree): *Rocky Mountains*

East of the Rocky Mountain Trench we come across the Canadian Rockies. Made up of four major ranges of mountains – the Park, Border,

Kootenay and Front Ranges – the "Rockies" extend from the US border to northern British Columbia. Along this rocky spine – the Continental Divide – lie some of the most spectacular and heavily glacierized regions of Western Canada. The icefields of Waputik, Wapta, Freshfield, Mons, Lyell, Columbia, Clemenceau and Chaba straddle this divide, their numerous outlet glaciers nourishing rivers on both sides, in some cases feeding the headwaters of three ocean drainage basins. Mount Snow Dome on the Columbia Icefield is in fact a hydrological apex from which water is drawn to the Arctic Ocean via the Athabasca and Mackenzie Rivers, the Pacific Ocean via the Columbia River, and the North Atlantic Ocean via the North Saskatchewan River and Hudson Bay.

Nearby Mount Athabasca was first climbed in 1898 by J. Norman Collie (1859–1942), a distinguished scientist, mountaineer, writer and art collector, along with Rocky Mountain notables Hugh Stutfield (1858–1929) and Herman Woolley (1845–1919). They were the first people of European descent to set eyes on the Columbia

Icefield. Collie describes the event thus in a book written by him and Stutfield, *Climbs and Exploration in the Canadian Rockies*:

> The view that lay before us in the evening light was one that does not often fall to the lot of modern mountaineers. A new world was spread at our feet; to the westward stretched a vast ice-field probably never before seen by human eye, and surrounded by entirely unknown, un-named, and un-climbed peaks. From its vast expanse of snows, the Saskatchewan Glacier takes its rise, and it also supplies the headwaters of the Athabasca; while far away to the west, bending over in those unknown valleys glowing with evening light, the level snows stretched, to finally melt and flow down more than one channel into the Columbia River, and thence to the Pacific Ocean.

Even today, little is known about the Columbia Icefield's volume and bed topography, though a study of the icefield and its outlet glaciers – to determine these items and to begin measurements of the icefield's changes – has recently been started by the Geological Survey of Canada and the Parks Canada Agency.

The large icefields and outlet glaciers of the region represent the majority of glacier area, but the bulk of the glaciers in the Rockies are the small mountain type. Typically, these mountain-type glaciers are found in the cradle of cirque-shaped valleys, where they are often afforded protection from the sun and reliably nourished by wind-drifting snow and avalanches from surrounding slopes. The glaciers of this region historically provided great highways for explorers and surveyors. By the glaciers' grace, these explorers attained many prominent summits, mapped provincial boundaries and defined watersheds. Furthermore, the glaciers enabled these pioneers to consider what they could only imagine from the valley bottom: a way through the seemingly impenetrable barrier of the Rockies and the potential for water riches to support the development and sustainability of the nearby prairies.

Nahʔą Dehé (Dene): *River of the Land of the Nahʔą People*

One remaining significant region of glacier cover in the mountain West lies near the Yukon–Northwest Territories border, in the Selwyn and Mackenzie Mountains. The Selwyn Mountains, named after Alfred R. Selwyn, director of the Geological Survey of Canada (1869–95), are further distinguished as having a northern section that lies to the east of the Yukon–Northwest Territories border, and a southern section that straddles the same border, extending from the mining heartland of Macmillan Pass to the South Nahanni River headwaters. Notably, the Geological Survey of Canada and Parks Canada recently inventoried the glaciers of the southern section in order to assess landscapes and habitats related to Nahanni National Park Reserve, which the Parks Canada Agency was hoping to expand. Happily, the park now includes the contributing headwaters of the World Heritage South Nahanni River system, as well as their glaciers.

This area and the region north and northwestward to the Yukon-Alaska border is held by glacial geomorphologists* to contain the most extensive glacial record of any preserved on Earth. Here we witness impressions left by glaciers during multiple regional-scale glaciations beginning between 2.9 and 2.6 million years ago. Textbook examples of glaciation abound, with several ranges, such as the Ragged, Itsi, Rogue and Rifle, still exhibiting small icefields, mountain glaciers and cirque glaciers. The Mackenzie Mountains, lying in the rain shadow of the Selwyn Mountains, are, in essence, a 750-kilometre-long northern extension of the Rocky Mountains. Here you can find remnant mountain glaciers and cirque glaciers.

By this point in our journey, we can all agree that *a glacier is not a glacier is not a glacier.* Glaciers are complex expressions of cold and warm on the landscape; while they are cold in some places, they are the source of water in others. Their forms are as variable as the landscapes they are confined by, overrun or carve.

* Geomorphology: the study of the physical features of the surface of the earth and their relation to its underlying geological structures.

These Great Carvers are not static but alive – places of spirits. Perhaps, on your hike through the refined mountains of the upper Ottawa Valley, you will contemplate: *Did a glacier carve those lines into the rock? Or was this river valley a former drainage channel formed by the immense flows of water emanating from the margins of the retreating ice sheet?* Or maybe you will book a window seat on your next flight across a glacierized region, looking forward to pressing your nose against the airplane window, straining to catch, in the vastness of an hour, a glimpse of our geological and glacial history.

Measures and Metrics

Had they never invented the thermometer and the scales of temperature, we would still have glaciers to tell us the story of regional and worldwide warming and cooling.

Glaciers reflect both long-term and contemporary climate changes, so their measurement is key to differentiating external or "extraterrestrial" influences – the sun's energy coming into our atmosphere – and terrestrial influences – processes and controls on our climate system originating from the interaction of the Earth's land, sea and atmosphere. The audacious approach – What?! You want to measure a glacier? – was first proposed by Swiss scientist F.-A. Forel (1841–1912) in a landmark 1895 article entitled "Les variations périodiques des glaciers." With over 100 years of direct glacier measurements under our belt since that time, the formal approaches have adapted as technology has permitted. We are now able to measure with remote sensing – scanning the earth by satellite or high-flying aircraft – what we could once only imagine existed.

Glaciers are measured in a number of ways for different purposes. Contemporary glacier variations are documented most easily by measuring changes in their size (area and length). The relationship between these measurements and climate is complicated, though, since changes occurring higher up on any glacier will take years, sometime decades or even centuries, to make their way down to the glacier tongue. The mass balance, a concept you are already familiar with, provides a more direct indicator of glacier health by showing if a glacier is experiencing overall mass increase or decrease or maintaining an equilibrium.

We rely on long-term perspectives on glaciers to determine if what is happening now is within or outside of the natural range of what has happened in the past (natural variability). When taken together, contemporary and historical evidence has the potential to provide us with four key features of our changing environment: sign – we attribute plus to increasing or minus to decreasing change; magnitude – how much change; trend – is the change consistently greater or less than an average, or normal; and acceleration or deceleration – does the change occur faster and faster or slower and slower.

The long-term perspectives necessary to determine these four key features can be derived from evidence left behind in the rocks and sediment, or moraines, created when the glaciers pushed forward to, and retreated from, their intermediate and maximum extents. Stands of forest may have been pushed and damaged, and trees, now tree stumps, may have been be overridden by the ice. All material that makes up this evidence can be dated using various scientific techniques, which helps us to estimate where and how large the glaciers were, and when.

Other evidence that helps us to tell the stories of the glaciers comes from the glaciers themselves. Vaults of information are constructed and stored as layer upon layer of snow accumulates, essentially resulting in the glaciers acting as museums of past atmospheric conditions and climate. Like a tree whose internal rings reveal the nature of past climate and the air it breathed, each new layer of snow on a glacier seals in the air and its components on lower layers – the great icy archivist.

As this manifesto project was seeded, Bob Sandford expressed to me that "a glacier is time compressed." Indeed, he is right. Some glacier configurations, particularly those where ice is not transported by flow or sliding toward the margins of the glacier, preserve time. Yet others, whose beds are at melting point, can only provide a sliding window on the past because the records are being removed at the bed while new records are amassing at the surface. The oldest glacial ice exists in glaciers that are frozen to their beds, and thus those spots are ideal locations for extracting columns of layered ice, called ice cores, which help scientists to reconstruct the past.

Ice-core-derived evidence of Earth's climate fluctuations has been key to unlocking the degrees of change that the Earth has endured during and prior to the dawn of modern civilizations.

Ice cores several thousand metres deep from the ice sheets of Antarctica and Greenland provide the longest records, giving us perspectives on multiple glaciations and interglacial periods. Shorter cores drawn from dry, high-elevation regions, such as the summit of Mount Logan, reveal, in less than 200 metres of ice thickness, conditions throughout the Holocene to the last part of the last glacial period, which ended approximately 10,000 years ago. From these cores, atmospheric temperature can be inferred, atmospheric chemistry and dust can be measured, and mathematical models of the age/depth relationship can be calibrated. When augmented with oral history perspectives and documented volcanic events, an ice core becomes a valuable portal into the past – a time machine.

As glacier chemists explore deeper into the glacier, they trade the sharpness or clarity of what they find, for time. The deeper scientists look, the more ice is compressed and smeared outward to form ever-thinner annual layers. Here, the physical sample size, called *support*, is simply too large to capture seasonal or annual variations. Thus we must give up some clarity and accept decadal to centenary averages for the climate and atmospheric parameters being studied. Proxy records have dramatically changed what we understand about the past and the present and what we can predict or hypothesize about the future, but they do suffer from limitations, such as that relating to the resolution or sharpness of well-aged ice cores.

Along with historical proxy records, contemporary measurements are crucial in answering questions like: What are the glaciers doing now, and how does that relate to present climate conditions? Canada and other nations committed to the United Nations Framework Convention on Climate Change (UNFCCC) contribute their measurements to a worldwide coordinated effort to observe changes in global climate. This is where the measurement of glacier mass balance comes in. Glaciologists measure glacier mass balance by using both ground and remote sensing techniques. In Canada, we take measurements over a network of reference glacier observing sites in our mountain West and Arctic Islands. Measurements of the snow depth that has accumulated on the surface after the winter season is complete are converted to a water equivalent using measurements of snow density. The results of this conversion give glaciologists the data with which to estimate the *winter mass balance* that has nourished the glacier.

Next, measurements of the remaining snow (now called firn – see page 25) after the summer melt season, and the amount of ice that has melted after the snow leaves the surface, allow us to estimate the *summer mass balance.* Taking the winter and summer balances together, we can compute the corresponding *net mass balance* for that *mass balance year* – in the northern hemisphere, usually between late August and early October, when nourishment begins, to the same time in the following year, when melting comes to an end. Not always a straightforward process – complications like the formation of flow fingers can make the estimation of winter and summer mass balances very difficult. Only with specialized expertise can one come to reliable estimates of glacier mass balance.

In the presence of such immense, slowly moving rivers, one is apt to forget the fullness of an hour and the briefness of life – but glaciers are in many instances so large that we could not conceive of them until the advent of remote sensing. In its earliest form, over a century ago in Canada, remote sensing meant standing atop a mountain peak with a camera and glass plate emulsion. This was followed by aircraft-supported photography, such as the oblique images of Bradford Washburn and Austin Post. Only a few decades ago would cameras in Earth's orbit begin providing nearly global surveillance at regular intervals.

Glaciologists employ remote sensing techniques so we can sample many glaciers, not just the reference mass balance glaciers. We need to define the state and evolution of Canada's glaciers over broad regions where they have an impact on water resources and other natural and human systems. It is simplest to survey glacier length and area from space with a satellite sensor that is sensitive to the optical spectrum. The resulting pictures shown as a time series do tell us how the glaciers are reacting to changes in mass balance, but they tell us little about the seasonal and annual mass balance changes we use to infer the weather and climate conditions that caused them. Moreover, optical methods using the visible spectrum are limited by cloud cover and the solar nighttime.

To estimate the annual net mass balance remotely, there is an ambitious program of surveying glaciers using aircraft fitted with a specialized scanning laser instrument that provides, when repeated annually, measurements of the surface elevation change of a glacier. Coupled with estimates of surface density, densification and motion, these *repeat laser altimetry* measurements can be used to estimate mass balance changes. Soon we will be able to get the same results – day, night and all weather – with a radar altimeter satellite called CryoSat. Glaciology and remote sensing research scientists are now working to develop the algorithms that will allow the data collected by this satellite to tell us about snow accumulation and ice melt; such a feat would allow us to map where the glacier surface is ice, and where it consists of snow, firn and areas of refrozen percolating water.

One other important remote sensing method glaciologists use to determine glacier mass balance over large glaciers or glacier systems involves a technique called *radar interferometry.* Using radar satellites, such as Canada's Radarsat 1 and 2 or the European Space Agency's Envisat, we measure glacier motion. Essentially, the arrival time of radar signals reflected from a satellite that is imaging a glacier surface at one time are compared against signals reflected from almost the exact same spot at a later time, which allows us to determine the amount of motion that has taken place during the interval between the two. The same thing can be done using two satellites, or one satellite with two radar systems, that are a known distance, or *baseline*, apart. Since it takes time for a single satellite to return to the place where it originally sent the radar signals, these later techniques using two satellites or one with two radar systems allow measurements over smaller time intervals, which is important when other fast-changing properties of a glacier also influence the radar response. Using these measurements of surface motion in conjunction with other factors, glaciologists apply the principle of mass continuity (see page 19): Is the ice flux from above the equilibrium line replacing the ice being lost below the equilibrium line?

Glacier mass balance is considered by the World Meteorological Organization's Global Climate Observing System to be an *essential climate variable*, with glaciers themselves having been referred to more casually as Climate Change Stations. First, unlike the climate stations located near urban heat islands and aerodromes, glaciers are generally removed from local anthropogenic influences but are subject to regional and global ones. Second, mass balance is a measure of what nourishes the glacier – water – versus what melts the glacier – heat – and so provides a robust, integrated measure of water and heat flux at Earth's surface. So valuable is this simple measurement that, if made over a sufficiently long period, climate signals can be distinguished from weather-generated noise.

Canada's Glacier Stewards

In the end, we will conserve only what we love.
We will love only what we understand.
We will understand only what we are taught.

— Baba Dioum, speech to the International Union for the Conservation of Nature, 1968

In Canada, the main government organizations involved in glaciological research are Natural Resources Canada and Environment Canada. Coordinated glacier-climate observations and assessments at Natural Resources Canada are conducted under an initiative called "The State and Evolution of Canada's Glaciers." This initiative delivers a "National Glacier-Climate Observing System" for Canada and conducts related freshwater flux research. Its outcomes are enabled by an integrated program of research and monitoring conducted by Natural Resources Canada's Geological Survey of Canada; Geomatics Canada – Canada Centre for Remote Sensing; Geomatics Canada – Mapping Information Branch; other government departments such as the Parks Canada Agency; and numerous university partners.

Research is conducted on i) the distribution and quantification of glacial mass change; ii) detecting change and what causes it, in relation to factors that influence climate and weather patterns; iii) water-resource and sea-level change and future potential change; and iv) developing better tools to observe glaciers, with less uncertainty and greater regional significance. Furthermore, baseline data and expert advice are provided to the environmental and natural resource sectors (protected areas, proposed and established mining and hydro power projects). Under Canada's obligations to the UNFCCC, data and assessments also contribute to the World Meteorological Organization's Global Climate Observing System; the United Nations Environment Programme's Division of Early Warning and Assessment; the United Nations Educational, Scientific and Cultural Organization's (UNESCO) International Hydrological Programme; and the Intergovernmental Panel on Climate Change.

Together with our colleagues in other nations, Canada's glacial stewards work to listen to the stories that glaciers tell and interpret them in the current context of a phenomenon that is broadly considered the largest and most serious challenge to be addressed by our society - our changing climate.

Becoming Water: Beacons of Climate Change, Changing Climate

Your acceptance of the risk of being too slow to
recognize change versus the risk of chasing
phantom blips should help you
decide which to give more credence to.

— Charles Franklin, Pollster.com, May 2, 2007

Because glacial mass is an integration of the long-term variability of precipitation, mean temperature and cloud cover, changes in glacial mass are considered among the most robust indicators of climate change.

In the work I contribute to – the quest to listen to the stories glaciers are telling us about past and current climate changes – I have the great privilege of not only working with many colleagues in Canada but also my contemporaries and elders from many nations. Notably, we have for numerous decades been influenced by the work and teachings of Wilfried Haeberli, the former director of the World Glacier Monitoring Service, whose coordinating office is based in Zurich. The following summary by Professor Haeberli, from his contribution to a book on glacier science and environmental change, should all sound somewhat familiar to you by now. As we continue through this chapter of the stories of glaciers, we will revisit some of these ideas. Professor Haeberli sums up some of what we have learned since the dawn of modern glaciological investigation more than a century ago:

> In 1895, it was envisaged by F.-A. Forel of Morges, Switzerland, that glacier observations would be the key to the question about the global uniformity and terrestrial or extra-terrestrial forcing of past, ongoing and potential future climate and glacier changes. Since then, various aspects involved have changed in most remarkable way – in particular that concern is growing that the ongoing trend of worldwide and fast,

> if not accelerating, glacier shrinkage at the 100-year time scale is of a non-cyclic nature – that there is hardly a question any more of the original question put forth by Forel – that under the growing influence of human impacts on the climate system, dramatic scenarios of future developments, including complete deglaciation of entire mountain ranges, must be taken into consideration – that such future scenarios may lead far beyond the range of historical/Holocene variability and most likely introduce processes without precedents in the Holocene.

The worldwide effort to measure contemporary glacier mass changes under the coordination of each participating country's World Glacier Monitoring Service office has resulted in the development of extremely valuable perspectives on our changing climate. When compared to indirectly estimated changes over several millennia, concern is growing that the ongoing trend of worldwide glacier shrinkage, at the century time scale, is of a non-periodic nature – in other words, is not a cycle.

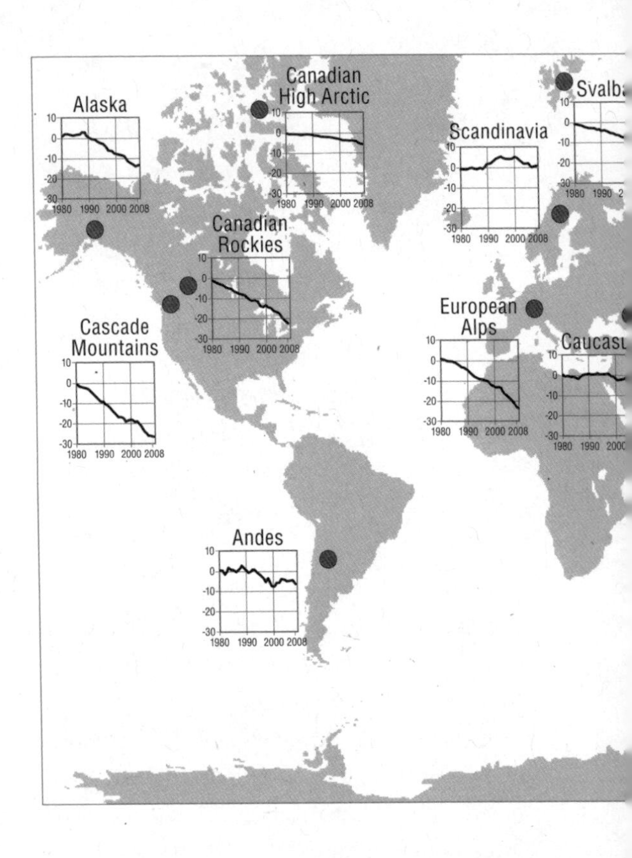
Alaska
Canadian High Arctic
Scandinavia
Canadian Rockies
Cascade Mountains
European Alps
Andes
10
0
-10
-20
-30
1980 1990 2000 2008

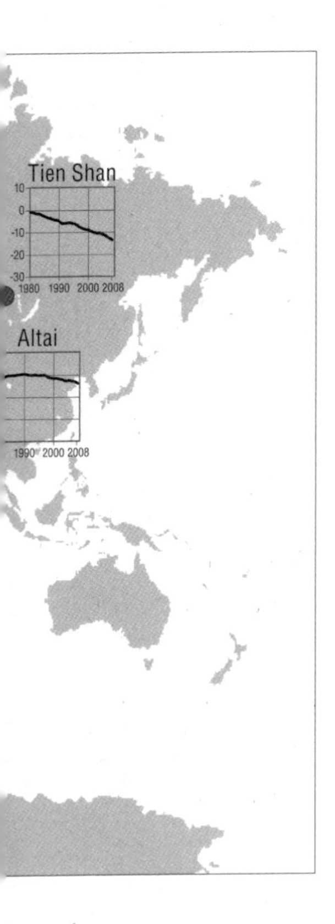

FIGURE 11: Glacier mass balance trends around the world, expressed as glacier thickness change in metres of water. (Adapted from World Glacier Monitoring Service data archive and WGMS-Canada/Natural Resources Canada)

Glacier shrinkage appears to be occurring quickly, particularly in continental climates, and accelerating in many regions. Where glacier mass increases have been observed, it has been determined that under the influence of warmer air temperatures, such sites, which are near the sea at high latitudes, have been receiving more nourishment that, for a period of time, will offset the loss of mass that happens during summer melting. So what appears to the casual observer to be contradictory evidence – glaciers gaining mass despite warming of Earth's atmosphere – is in fact complementary evidence, since warming can manifest itself in different ways for glaciers in different climates and regions. As we know, fluctuation of a glacier's mass is not just the result of melting processes on the lower reaches of the glacier in summer.

The upper expanses of a glacier are our future. They are places where messages are being written to our descendants, evidence of what we have and have not done to be more mindful of the slow infrastructure that is our air, water and land. Think about where you were *not* when the first formal observations of glaciers began in Canada – around 1896 – or in the European Alps in the 1700s. As we've already seen, drilling down into the details of such ice masses reveals much information about the past. Analysis of the ice cores brought to the surface yields evidence not only of the natural variability before humans started influencing the atmosphere with large-scale natural-resource harvesting and agriculture, but also of the effects of the modern industrial period.

Listening to the Signals through the Noise

And when it melted out of the ice, would it then just sublimate back into metaphysical space, leaving human time and scientific measurement behind?

— *Thomas Wharton,* Icefields *(1995)*

In examining data records of any sort, it is common to determine whether there are trends, cycles and shifts in the average values that are distinguishable from random noise. As an exercise, let's examine some data – not necessarily about climate or the mass balance of a glacier, but data of any sort that varies with time and so forms a *time series*. When you refer to the figure on page 97, you see that time marches along the horizontal axis, increasing to the right. The magnitude of whatever the data represents varies vertically, with values increasing upward or decreasing downward. This synthetic data exhibits several important features: *noise*, *cycles*, *trends* and *shifts in the average*.

Noise is random variation that, if you were to sample it over and over again, then add up all of those sample values and take their average, you would get a straight line that would not vary with time. We have ways to reduce the effects of noise, but for now it is just important to understand that it exists. Cycles represent repeated patterns, and trends represent gradual but persistent change through time. Shifts in the average represent sudden or temporary occurrences, where conditions before and after a reference point are demonstrably different – a step change. My hope in sharing this statistical information is not to overwhelm, but rather to illustrate just some of the considerations involved in the discussions around climate change and its effects on our ecosystems.

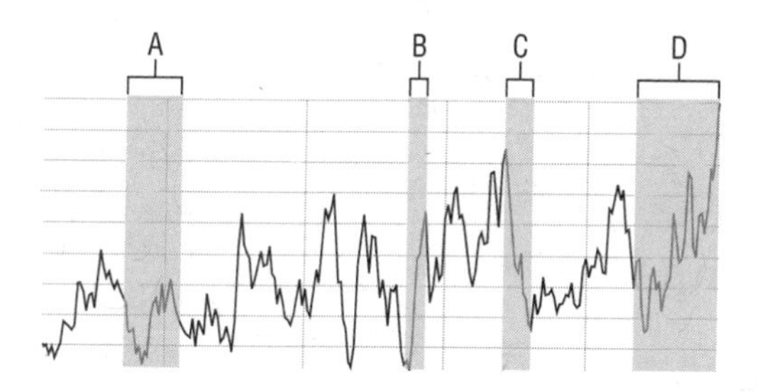

Figure 12: Presented here is a synthetic data series that could represent air temperature at the summit of Mount Waddington or sea level change at Clyde River or the depth of snow on top of a glacier in Norway. Time marches from left to right along the horizontal axis. There are trends up and down and shorter- and longer-period cycles, all superimposed. Using two pieces of paper as a mask, "zoom in" on the "windows", or anywhere you like, and decide what the data is telling you. Depending on how much data you are looking at, some of the patterns in it could be called "noise." Finally, begin to "zoom out" gradually and decide if your first conclusions are the same. The important lesson is not to jump to conclusions using a time series of data that is too short. Remember also the importance of knowing what the vertical and horizontal scales are, so that you can begin to recognize what is noise and what is "signal." As a further exercise, revisit Figure 8 on page 48.

Still referring to the figure on page 97, note that there are several time windows in the time series: A, B, C and D. Take a moment to examine the nature of the data you see. What is it telling you? As you can see when closely examining the entire series, the features are all overlaying each other – there is noise superimposed on trends, trends within cycles, cycles within yet bigger cycles, and all of it within an overall trend. Different amounts of time are required to recognize each feature and, as we look at more and more of the figure, our conclusions may be very different than if we were just looking at one window. Generally speaking, for instance, cycles tend to be natural phenomena, such as how Earth orbits the sun and is subject to long-period variations in the amount of solar radiation that reaches an area. Persistent trends suggest anthropogenic influences, though again these may be superimposed on or even hidden within larger cycles.

So, with all this variation, what is *normal*? The World Meteorological Organization and the Meteorological Service of Canada consider that a *climate normal* is established after 30 years of observation. By continuing to gather data beyond that time, climatologists can begin to determine whether the normal is changing and how it may be changing. It's important, however, to first recognize the difference between *weather* and *climate*. Many people do not, as evidenced by statements about chilly summer days: "So much for global warming, eh?" Or circumspect statements about modelling future climate trajectories: "If they can't predict what the weather will be in five days, how can they then predict what the climate will be in 50 years?" As so well described in Andrew Weaver's book *Keeping Our Cool: Canada in a Warming World*, it is in fact altogether easier to predict climate than weather

Simply stated, weather is what you experience now, today, where you are. Climate is what you expect. In examining data to extract meaningful climate-related statistics, climate scientists, glaciologists and hydrologists are looking for signal and, in some cases, trying to find it amid a great deal of noise. At certain time scales, usually smaller, weather is considered noise. Now, say we were to expand time out to consider several glacial cycles. In that scenario, even generational cycles and trends that we can comprehend – along with climate patterns that affected the establishment of modern civilizations – would be considered noise. So the question becomes: What should the size of our time window be so that we may adequately characterize natural variability (and therefore be warned that something may be developing beyond a range that human and natural systems can withstand or adapt to)? This is a difficult but crucial question in deciding what we should be mindful of when examining time series data.

I would suggest to you that the time scale of importance, as it concerns climate variability and climate change as a phenomenon in our human context, is the time during which modern civilization has unfolded (about 7,000 years) and continues to evolve. We need to be mindful of the scales of establishing human infrastructure and adaptation to terrestrial conditions – people plan, make decisions and build infrastructure over 10 to several hundred years for instance – together with those of slower, natural infrastructure such as the water cycle and climate.

We have already introduced the need to examine not only past and current changes but also possible futures. The realm of predicting climate and ecosystem/hydrological functioning is vast and complex, requiring the skills of highly experienced climate system physicists and those who characterize ocean, atmosphere and landscape change. An important approach toward assessing the predictive power of climate models is to see if they can adequately predict what has already happened.

To varying degrees, climate models are capable of incorporating the external or "extraterrestrial" forcings – the sun's energy coming into our atmosphere – the effect that atmospheric constituents, both natural and human-caused, have on terrestrial conditions. These include natural dust, water vapour, forest fire charcoal, volcanic ash, acid aerosols and greenhouse gases. By running the models with and without the factors specifically known to be the result of human activities, for instance *excess* greenhouse gases, and comparing the results with actual observations of air temperature, modellers can do two things: assess the predictive power of their models, and determine both the timing and the magnitude of the changes induced by natural versus anthropogenic causes.

Human Vulnerability and Adaptation

A man who keeps company with glaciers comes to feel tolerably insignificant by and by. The Alps and the glaciers together are able to take every bit of conceit out of a man and reduce his self-importance to zero if he will only remain within the influence of their sublime presence long enough to give it a fair and reasonable chance to do its work.

— *Mark Twain,* A Tramp Abroad *(1880)*

Having robust models of our climate future is a fundamental part of enhancing our resiliency to the impacts of a changing climate. Armed with information and assessments of its uncertainty on how warm and wet it will get generally, and how dry in some regions specifically, will foster our ability to plan, act and adapt.

While glaciers participate in this process of relentless change, in part, through the terrain they carve, gaining resilience or vulnerability depending on the topography, we humans are, to a significant degree, responsible for our own, global resilience or vulnerability in the face of such unprecedented change. This is transformation beyond what we are normally able to cope with or perhaps even comprehend, because the only change we have experienced thus far has been over intervals that exhibit a well-defined average condition, and fluctuation around that average. The capacity to withstand conditions outside of this normal average is termed *adaptation*, and we become vulnerable when conditions exceed our capacity to adapt. The recognition that we are running up against the limits of our coping range encourages environmental monitoring, and this in turn plays a crucial role in addressing a societal phenomenon termed the *adaptation deficit*.

Environment Canada's lead expert on climate change adaptation, Ian Burton, coined and defined the term *adaptation deficit*, which essentially describes the behaviour related to not recognizing or fully acknowledging the resounding evidence that our climate system and related services are headed beyond our coping range. Some of the reticence that is key to being in a state of adaptation deficit is part of human nature; denial, fed by any evidence that appears to suggest the contrary, is our emotional escape hatch. Perhaps we deny the obvious because we simply feel we have too much to worry about already. Being in such a state of adaptation deficit, however, diminishes the individual and collective pool of emotional resources that we need in order to deal with decision-making in the face of what, quite frankly, is some pretty heavy stuff.

Recent work (2010) published by NASA's Goddard Institute for Space Studies, the group responsible for the analysis of the instrument-based records describing global surface temperatures, concludes that:

> ... contrary to popular misconception, the rate of warming has not declined. Global temperature is rising as fast in the past decade as in the prior two decades. We conclude that global temperature continued to rise rapidly in the past decade, despite large year-to-year fluctuations associated with the El Niño–La Niña cycle of tropical ocean temperature. Record high global

temperature during the period with instrumental data was reached in 2010.

Not only are the consequences of warming beyond our adaptation limits dire, the climate system and its related services are complex and the subject of much continued and necessary study. We are learning along the way. It is rather like building public transportation infrastructure in advance of the population that will use it: if you provide people with honest and rigorous information they can use to make decisions, *they will come*. What makes the task considerably more difficult, though, is that the cases "proving" or "disproving" human-induced climate change are often presented as one or the other – anthropogenic or natural – when in fact both are occurring simultaneously, sometimes reinforcing each other, sometimes neutralizing each other. Long observation and historical records allow us to sort this out and provide the basis for sound adaptive and risk management, arguably at a measured pace that does not swing us into periods of instability that create societal and environmental chaos.

FIGURES 13a and 13b: The Bow Glacier, photographed in 1901 by the Vaux family and again in 2001 by Henry Vaux Jr. Both photos were taken in late summer of their respective years. (Vaux Family fonds, with permission, Whyte Museum of the Canadian Rockies) (with permission, Henry Vaux Jr.)

Collaboration and Direction

The pervasive impacts of weather and climate in people's livelihoods, economic sectors and the natural environment have made adaptation to climate change everybody's business.

— Ian Burton, Professor Emeritus, Simon Fraser University, December 2005

Ian Burton's seminal work in climate change adaptation makes the case that climate change needs to be factored into decision-making at all levels and across all sectors, including the manner in which issues related to gender equity and the eradication of poverty are tackled. Climate change is finally being recognized as a poverty and equity issue by international consensus, and whatever the balance of its causes, we must recognize that the burden of adapting to the impacts of globally synchronous warming will be carried by the world's poorest nations and most vulnerable people. In that light, we must make an important distinction – and adjust our actions – between *responsibility* for climate change impacts and *concern* for natural disasters. Burton notes: "The grim prospect is that with climate change the adaptation deficit as a whole is set to grow significantly larger." We must find new ways to look at the issues. Greenhouse gases and "pollution," for instance, are so different that narrowly talking only of mitigation through technological innovation will only deepen the adaptation deficit.

In the face of an increasingly complex world, old models of domination and conquest have to give way to those that foster diverse partnership and collaboration. Every interest has part of the story. In the science policy arena, however, a disturbing trend has emerged over the last half-century. At a Canadian conference on domestic and international water security held in 2010, Ralph Pentland, an expert on water policy issues (in particular those concerning ecological integrity and the management of transboundary waters under the stress of climate change) gave a telling keynote address. In it, he emphasized a need for sound scientific judgment; the kind that results from the two-way flow of information and the adoption of policy that would help citizens adapt, at least in part, to climate changes that are largely beyond their individual control. Too often, Pentland lamented, science results are cherry-picked to reflect particular agendas or points of view.

It is fortunate, then, that we have the United Nations Intergovernmental Panel on Climate Change (IPCC), whose goal is to provide consensus information on the causes and current state of climate change and its impact, and advice on how to adapt to and mitigate the forces of change. While not without the troubles of any large organization, the IPCC does not fund research, as has been put forward by some looking to discredit the organization's motives. The IPCC is one of the instruments used to deliver on the commitments of signatory nations to the United Nations Framework Convention on Climate Change (UNFCCC). As it concerns climate change adaptation, the UNFCCC did foster the development of the Global Environment Facility, a financial instrument that, in part, supports adaptation in developing countries.

Troubling, though, are recent proclamations that we might prosper during an era of climate change. There is certainly evidence that food production in some regions may be enhanced by an increase in air temperature, only to be confounded in some cases by regional desiccation of the available water, or drought. While these concepts are worth exploring, the overarching principle that should guide our policy-making and our lives is *What if?* – particularly as we strategically migrate away, as Australia has, from communicating using the words *climate change* to instead saying *a changing climate*.

The UNFCCC is founded on creating a consensus approach that would foster:

> … the stabilization of greenhouse gases in the atmosphere at a level that would prevent anthropogenic interference with the climate system. Such a level should be achieved within a time frame sufficient to allow ecosystems to adapt naturally to climate change, to ensure that food production is not threatened and to enable economic development to proceed in a sustainable manner.

To aid this, the *precautionary principle* is applied: reductions in excess greenhouse gases and other pollutants would be a good thing, regardless.

We have to recognize that damage has already been done to the atmosphere and that there will be difficult work ahead in a world that tends to orient itself toward short-term results. And even where a longer view prevails, measures of success are often cast in relative terms, as opposed to absolute reductions in exponentially increasing physical phenomena or societal conditions deemed to be intolerable.

It is often said that we are small in the presence of nature. But this does not mean we are not capable of exerting negative influences on nature. Some camps purporting to refute the existence of anthropogenic climate warming use the argument of our diminutive existence in the presence of nature to launch such leading questions as: How could we be so supercilious as to suggest we have any such influence on climate? An interesting, if not malevolent, twist in logic. The findings of the IPCC support that the balance of the evidence strongly suggests that human-caused climate change and other anthropogenic influences have been warming and will continue to warm the atmosphere globally at a rate, and to a magnitude, well beyond what is suggested in proxy measures of natural variability.

So consider once again Figure 8 on page 48. Examine its data and ask mindful questions just as you would of people presenting any data, such as, say, investment returns or insurance coverage. And with just such open-minded rigour, you would be challenging what may well be spurious analysis. If such analysis, if deemed invalid, were to form the basis for public policy, it could easily result in untenable but all too foreseeable outcomes far removed from anything ecologically or socially sustainable.

I believe it is within our power as a society to adapt and innovate to ultimately reduce the harm we exert on the planet's generally large, but regionally finite, capacity to buffer. A good example of just such successful coping is the "ozone hole" issue and the 1987 Montreal Protocol on Substances that Deplete the Ozone Layer. The protocol was the successful outcome of long, deliberate work by public science collaborating with industry and the public to eradicate ozone-depleting substances from our temperature-control appliances and aerosol propellants. Another, related example would be the UV index that is now part of the flow of our daily lives. If we are to approach climate change today with the same rigour that we so successfully applied back then to those equally large, and at the time equally contentious, atmospheric problems, we must continue to ask that policy objectives are informed with rigorous science and consensus. And of course, we need to recognize and support the quiet but crucial work of scientists, both physical and social, that produce that science and consensus.

Because there are glaciers in so many of the regions of the world where we do not have heat and water flux measurements, the story told (and recorded) by our Canadian glaciers is made that much more noteworthy. But it is a complex story, and a changing climate is a complex phenomenon. Together, we must listen to the stories of places nearby or far enough away that we may never visit. And together, we must demand that we all acknowledge that these places are vulnerable and of vital importance to the ecosystems we all rely on for our survival.

Epilogue

But it will go from more than ever
to not enough in no time

—Ani DiFranco, "78% H_2O" (2006)

It is my sincere wish that an understanding of glaciers, changing climate and water from the points of view presented in this short work should engender in all of us sensibilities that author Stewart Brand describes as "serving the long view and the long viewer." We should reward patience, and work to influence innovative solutions with, not against, our competitors. While doing so, we are wise to be mindful of Earth's mythic depths and unknowns, for we must leverage longevity. Our future will demand that we keep records, and in doing so we will foster an interest in renaissance, history and legacy. If we can do this, then we will be well on the way to minding the safety and life cycle not only of engineered infrastructure but also of slower natural infrastructure such as water, climate and ecosystems.

Things You Can Do, Questions You Can Ask Yourself

- Explore upstream to the source of your water. Learn about the ecosystems that keep it whole, and tell that story to others.
- Share your knowledge about Great Carvers. Check out one of the most superb graphical efforts to describe textbook glaciation on the web and share the link with friends on Twitter or Facebook: http://en.wikipedia.org/wiki/Glacier.
- If you can, visit a glacier. Walk on it, fly over it and fall in love with it. Be the water within, yearning for the water without.
- If you can't visit glaciers, imagine them. Use the Internet to help spark your imagination by searching terms you see in this manifesto, like "glacier calving" or "ice island" and see what videos, images or news stories you can

find. Listen to the stories they tell you of local, regional and global changes.

- Imagine the grief you will feel when something you love leaves you, and then love it more fully and completely.
- Recognize climate change symptoms during the flow of your own life, work and play. Are you enjoying a prolonged, more pleasant autumn at the cottage, or is your effort to open up the cottage in spring frustrated by cold and heavy squalls of snow? Is the approach to your favourite alpine snow and ice ascent less straightforward than it once was? How do your heating and cooling costs compare to a decade ago?
- Don't pass over the stories in newspapers or magazines about climate change and our ecosystems. Read them so that you are informed on all sides of the debate, then participate when topics are discussed. Dialogue is fundamental to both our and the earth's plight.
- You are aware that *weather* and *climate* are different but related. Share that knowledge.

- Recognize that changing climate is not an isolated issue amongst the list of many things your emotional resources might consider, but that its manifestations touch on nearly every other concern we might have, and as such, have to all be considered together.
- Explore data available from climate information services and early warning systems. Contact your community leaders to demand that such services and systems be adequately supported so that you can rely and make large decisions based on them.
- Ask planners and decision-makers in your community whether they use information on climate change in formulating policy. Do they have a source-water protection program or partnerships that promote planning for climate change? Are they increasing the resiliency of your community by being better prepared for severe weather, rising sea levels, drought and flood?
- Be mindful of your water, even if you are in an area of water riches – infiltrate your runoff, for example, from the roof of your

home, directly into the ground, not into drainage ditches and rivers.

- Try as best you can to conceptualize time, and recognize, as poet Don McKay does, that "place is where memory begins, and memory is how we place ourselves in the time in which we live."
- When you feel satisfied with your role and place in the world, consider Leonard Cohen's words: "There is a crack in everything ... that's how the light gets in."

Finally, since it was ecological historian and water expert Robert Sandford who encouraged the telling of this story, let me end with an invocation of his. Of all the things in this book, you should take with you and retell this thought:

> Dropping out of the sky from less blessed places, visitors from abroad instantly see what we sometimes forget. What makes Canada utterly unique in the world's imagination is that water exists plentifully here, in all its remarkable forms. Without abundant water we would be a very different

nation and a very different people. In the world's imagination, ours is a land of ice and snow, lakes, wild rivers, glaciers and icefields. That doesn't have to change, but we do. ...To fully understand emerging problems associated with water quality and availability in Canada, we have to return to our cultural headwaters to rethink what water means to us.

World Glacier Monitoring Service

Nations that contribute to worldwide coordinated glacier-climate observing under the auspices of the World Glacier Monitoring Service (not in alphabetical order, but grouped roughly by continental region):

Argentina
Bolivia
Chile
Colombia
Ecuador
Peru
Mexico
US
Canada
Denmark
Iceland
Norway

Sweden
Austria
France
Germany
Italy
Switzerland
Russia
Poland
Kazakhstan
Uzbekistan
Kenya
Tanzania
Uganda
Iran
Nepal
China
Japan
New Zealand

Appreciation

You reflect back to each other,
and it grows.

— Leslie Feist, musician, songwriter, singer,
collaborator on the Anthony Seck documentary
Look What the Light Did Now *(2010)*

My good friend Robert Sandford said to me: "[Mike], I believe that it is up to people like [all of] us to find the language, create the images and imagine the solutions that will allow us to break out of the vicious circle that threatens public health by threatening our landscapes and water sources."

It is this single expression that motivated me to get this manifesto started – at least to the point where it could be nurtured along by the superb advice and editorial force of Rocky Mountain Books' Joe Wilderson and Meaghan

Craven. Further editorial insight came from Shauna Rusnak, who was instrumental in the throws of final editing – thoughtfully influencing the narrative flow, insuring clarity and working with the RMB team to get the project landed. Shauna, I am most grateful. Inspiration and motivation also stemmed from Don Gorman's passion for mountain culture and historical and ecological entrepreneurship. Thank you, Don, for the opportunity to make a contribution.

I have benefited from numerous discussions, some that took place long before writing began and many more recent ones that have helped to crystallize the story and develop the imagery and character used in its telling. Thank you, Andrea Banks-Demuth, Greg Brooks, Ross Brown, Guy Buller, Dave Burgess, Sasha Chichagov, Graham Cogley, Seléna Cordeau, Margaret Demuth, Caroline Duchesne, Art Dyke, David Fisher, Don Forbes, Barry Goodison, Robert Gurney, Wilfried Haeberli, Phil Hill, Jocelyn Hirose, Gerald Holdsworth, Marika Hurko, Thomas James, Georg Kaser, Danielle Leavoy, Laura Lynes, Tony Maas, Shawn Marshall, Lynn Martel, Margo McDiarmid, Carole McLachlan,

Liz Morris, Kathy Muldoon, Scott Munro, Jamie Oliver, Andrew Nikiforuk, Simon Ommanney, John Pomeroy, Andrew Rencz, Bob Sandford, Vi Sandford, David Scott, John Sekerka, Wendy Sladen, Gwenn Story, Steve Wolfe, George Werniuk, and Michael Zemp.

I have the privilege of serving the people of Canada in my work with the National Research Council, Environment Canada and Natural Resources Canada, and in association with the Parks Canada Agency and Statistics Canada. This work has as much allowed me to measure and inform as it has to explore and be taught. The World Glacier Monitoring Service, and my role as the national correspondent for Canada, assisted me in articulating the data basis that tells the Canadian and global story. All factual errors are my own, as are the opinions expressed or implied.

The maps depicting the distribution of Canada's former covering of ice were inspired by those originally created by *Canadian Geographic* cartographer Steven Fick. Except as noted, the balance of the illustrations and technical line art were inspired by and articulated as a result of my collaboration with Danielle Leavoy of Arnprior, Ontario, and Chyla Cardinal of Rocky Mountain Books, Calgary. Thank you, Danielle, for reflection, friendship and seeing my mind's eye. Thank you, Chyla, for shoehorning it all together and making it look, read and feel fantastic; and of course, for creating another amazing manifesto cover!

The Arnprior Book Store and Bonnie Jane's Scones, "where friends gather," provided a sunny corner where I was able to write and be nourished by many good things. Mojtaba Nojoumi of Arnprior's Photo Max assisted me in recovering stories of the landscape I had previously photographed, all of which helped with the visualization required for the project.

Writing this manifesto also permitted me to recall and get back in touch with many an old friend with whom I had shared a rope, snow cave or hut while climbing and skiing in the mountain West, including Terry Beck, Steve Bertollo, Brant Hannah, Chic Scott, Helen Sovdat, Murray Toft and Danny Verrall. I miss you still, Alison. It is in the spirit of those shared experiences, small and large, that I hope this book will evoke a sense of honouring memory and history and serve our outlook to the future.

Last but not least, this manifesto would not have been possible without the support and encouragement provided by my family: Margaret, my partner in shared encounters with the wilder shores of many things, for good counsel, her razor-sharp sense of humour and sensitive way in which she processes the world; my daughters, Andrea and Mary, both beautiful women; and Matt, for bringing together two oceans.

Bookshelf

Sources

Allaby, Michael, ed. A Dictionary of Earth Sciences. 3rd ed. Oxford: Oxford University Press, 2008.

Alley, Richard B. *The Two-Mile Time Machine: Ice Cores, Abrupt Climate Change, and Our Future.* Princeton, N.J.: Princeton University Press, 2000.

Barber, Katherine., ed. The Canadian Oxford Dictionary. 2nd ed. Oxford: Oxford University Press, 2004.

Burton, Ian. "Adapt and Thrive: Options for Reducing the Climate Change Deficit." *Policy Options* 27, no. 1 (December 2005–January 2006): 33–38.

Cogley, J. Graham. "Mass and Energy Balances of Glaciers and Ice Sheets." In *Encyclopedia of Hydrological Sciences.* John Wiley & Sons, 2006. Abstract available at http://is.gd/EpP5sw (accessed 2011-08-30).

Collins, D.N. "Suspended sediment and solute delivery to meltwaters beneath an alpine glacier." In D. Vischer, ed. "Schnee, Eis und Wasser alpiner Gletscher." *Mitteilung* nr. 94 der Versuchsanstalt für Wasserbau, Hydrologie und Glaziologie an der ETH Zurich (1988): 147–162.

Demuth, Michael N., et al., eds. *Peyto Glacier: One Century of Science.* Ottawa: Environment Canada, 2006. Full text (zip-archived chapter PDFs) available at http://is.gd/n4crFT (accessed 2011-08-27).

Dyke, A.S. "Late Quaternary Vegetation History of Northern North America Based on Pollen, Macrofossil, and Faunal Remains." *Géographie physique et Quaternaire* 59, nos. 2, 3 (2005): 211–262. Abstract available at http://is.gd/jq075z; full text at http://is.gd/4lZnNa (both accessed 2011-07-22).

Forel, F.-A.: "Les variations périodiques des glaciers: Discours préliminaire." *Extrait des Archives des Sciences physiques et naturelles* XXXIV (Genève, 1895): 209–229.

Fritzsche, Jeff. "Trends in glacier mass balance for six Canadian glaciers." Ottawa: Statistics Canada Fall 2010 Publications. http://is.gd/tp9SFN (accessed 2011-08-27).

Gadd, Ben. *Handbook of the Canadian Rocky Mountains.* 2nd ed. updated. Canmore, Alta.: Corax Press, 2000.

"Global Glacier Changes: Facts and Figures." Zurich: World Glacier Monitoring Service/United Nations Environment Programme, 2008. Full text (PDF) available at www.grid.unep.ch/glaciers (accessed 2011-08-27).

"Global Land Ice Measurements from Space (GLIMS)." Worldwide glacier-monitoring database maintained by the (US) National Snow & Ice Data Center, Boulder, Colo. www.glims.org (accessed 2011-08-27).

Haeberli, Wilfried. "Integrated Perception of Glacier Changes: A Challenge of Historical Dimensions." In Knight, *Glacier Science and Environmental Change*, 423–430.

Hansen, J., et al. "Global surface temperature change." *Reviews of Geophysics* 48, RG4004 (Dec. 14, 2010). Full text (PDF) available at http://is.gd/Komlc2 (accessed 2011-08-30).

"Improved Processes and Parameterization for Prediction in Cold Regions." Saskatoon: U. Sask. Centre for Hydrology, 2010. www.usask.ca/ip3 (accessed 2011-08-27).

"Intergovernmental Panel on Climate Change Fourth Assessment Report: Climate Change, 2007 (AR4)." Geneva: United Nations Environment Programme and World Meteorological Organization. Full text (PDF) available at http://is.gd/LHype8 (accessed 2011-08-27).

Knight, Peter G., ed. *Glacier Science and Environmental Change*. Mississauga, Ont.: Wiley-Blackwell Canada, 2006. Table of contents available at http://is.gd/vgluNF (accessed 2011-08-27).

Mayhew, Susan., ed. A Dictionary of Geography. 4th ed. Oxford: Oxford University Press, 2009.

Orlove, Ben, Ellen Wiegandt and Brian H. Luckman. *Darkening Peaks: Glacier Retreat, Science, and Society*. Berkeley: University of California Press, 2008.

Paterson, W.S.B. *The Physics of Glaciers*. 3rd ed. Oxford, UK, and Woburn, Mass.: Butterworth-Heinemann, 1998.

"Retreat of the Laurentide Ice Sheet." Animated change-visualization map by Illinois State Museum, Springfield, 2006. Catalogued at Digital Library for Earth System Education, http://is.gd/Q9g9jh (accessed 2011-08-27).

"Rivers of Ice: Canada's Glaciers." Poster map and pictorial, W. Blake et al., scientific consultants. *Canadian Geographic* 118, no. 7 (November/December 1998).

Solomon, S., et al., eds. "Working Group I: The Physical Science Basis, Section 9.7: Combining Evidence of Anthropogenic Climate Change, Figure 9.5." In "Intergovernmental Panel on Climate Change Fourth Assessment Report: Climate Change, 2007 (AR4)."

"State and Evolution of Canada's Glaciers." Workspace maintained by Natural Resources Canada–Geological Survey of Canada with allied government agencies and university partners. http://is.gd/SdchDh (accessed 2011-08-27).

"Torngat Mountains National Park of Canada." Parks Canada. http://is.gd/955zeS (accessed 2011-08-27).

Williams, Richard S. Jr., and Jane G. Ferrigno, eds. "Satellite Image Atlas of Glaciers of the World: North America." United States Geological Survey Professional Paper 1386J. http://pubs.usgs.gov/pp/p1386j/ (accessed 2011-08-27).

Further reading

Brand, Stewart. *The Clock of the Long Now: Time and Responsibility*. New York: Basic Books, 1999.

Cruikshank, Julie. *Do Glaciers Listen? Local Knowledge, Colonial Encounters, and Social Imagination.* Seattle: University of Washington Press, 2005.

Davidson, Peter. *The Idea of North.* London: Reaktion, 2005.

Descent into the Ice. Documentary film by Liesl Clark. PBS television (US) airdate February 10, 2004; first aired in Europe as *Mont Blanc,* 2003. Transcript linked from main page at www.pbs.org/wgbh/nova/mtblanc.

"Glacier." http://en.wikipedia.org/wiki/Glacier (accessed 2011-08-27).

Haig-Brown, Roderick L. *Measure of the Year: Reflections on Home, Family and a Life Fully Lived.* Victoria, BC: TouchWood Editions, 2011. First published 1950 by Morrow.

Hewitt, Kenneth, and Ian Burton. *The Hazardousness of a Place: A Regional Ecology of Damaging Events.* Toronto: University of Toronto Press, 1971.

"Kettle Moraine State Forest Geological History." Madison: Wisconsin Department of Natural Resources, 2009. http://is.gd/1ZKJ01 (accessed 2011-08-27).

Kolbert, Elizabeth. "The Climate of Man I: Disappearing Islands, Thawing Permafrost, Melting Polar Ice: How the Earth is Changing." *The New Yorker* (April 25, 2005): 56.

———. "The Climate of Man II: The Curse of Akkad." *The New Yorker* (May 2, 2005): 64.

———. "The Climate of Man III: Global Warming and the Bush Administration." *The New Yorker* (May 9, 2005): 52.

Morley, David. *Mandelstam Variations*. Todmorden, UK: Littlewood, 1991.

Phare, Merrell-Ann S. *Denying the Source: The Crisis of First Nations Water Rights*. Calgary: Rocky Mountain Books, 2009.

"The Program on Water Issues." www.powi.ca/index.php.

Sandford, Robert W. *Ecology and Wonder in the Canadian Rocky Mountain Parks World Heritage Site*. Edmonton: Athabasca University Press, 2010.

———. "Interpreters of Natural and Human History Ltd." www.rwsandford.ca (accessed 2011-09-12).

———. *Restoring the Flow: Confronting the World's Water Woes*. Calgary: Rocky Mountain Books, 2009.

———. *Water, Weather and the Mountain West*. Calgary: Rocky Mountain Books, 2007.

———. *The Weekender Effect: Hyperdevelopment in Mountain Towns*. Calgary: Rocky Mountain Books, 2009.

Sandford, Robert W., and Steve Short. *Water and Our Way of Life*. Fernie, BC: Rockies Network, 2003.

Stutfield, Hugh E.M., and J. Norman Collie. *Climbs and Exploration in the Canadian Rockies*. Calgary: Rocky Mountain Books, 2008 . First published 1903 by Longmans.

Weaver, Andrew. *Keeping Our Cool: Canada in a Warming World*. Toronto: Viking, 2008.

Western Canadian Cryospheric Network. wc2n.unbc.ca.

World Water Assessment Programme. *Water, A Shared Responsibility: The United Nations World Water Development Report 2*. Paris: UNESCO, 2006.

Wharton, Thomas. *Icefields*. Edmonton: NeWest Press, 1995.

Wiebe, Rudy. *Playing Dead: A Contemplation Concerning the Arctic*. Edmonton: NeWest Press, 2003.

World Glacier Monitoring Service. www.geo.uzh.ch/microsite/wgms/index.html.

Yorath, C.J. *Where Terranes Collide*. Victoria, BC: Orca Book Publishers, 1990.

About the Author

Mike Demuth hails from Calgary and has studied snow and ice in its various forms on land and water for the last 30 years with the National Research Council, Environment Canada and Natural Resources Canada as a glaciology/cold regions research scientist. Mike's attention to studying changes in Canada's mountain West and the Canadian Arctic was secured by his participation in a research expedition to Mount Logan in 1981. This book is his first public outreach manuscript on climate science, water and the stories that glaciers tell. He and Margaret live in Braeside, Ontario, and have two daughters and a granddaughter.

Other Titles in this Series

Digging the City

An Urban Agriculture Manifesto

Rhona McAdam

ISBN 978-1-927330-21-0

Little Black Lies

Corporate & Political Spin
in the Global War for Oil

Jeff Gailus

ISBN 978-1-926855-68-4

Gift Ecology

Reimagining a Sustainable World

Peter Denton

ISBN 978-1-927330-40-1

The Insatiable Bark Beetle

Dr. Reese Halter

ISBN 978-1-926855-67-7

The Incomparable Honeybee

and the Economics of Pollination
Revised & Updated

Dr. Reese Halter

ISBN 978-1-926855-65-3

The Beaver Manifesto

Glynnis Hood

ISBN 978-1-926855-58-5

Ethical Water

Learning To Value What Matters Most

Robert William Sandford
& Merrell-Ann S. Phare

ISBN 978-1-926855-70-7

The Grizzly Manifesto

In Defence of the Great Bear

by Jeff Gailus

ISBN 978-1-897522-83-7

Denying the Source

The Crisis of First Nations Water Rights

Merrell-Ann S. Phare

ISBN 978-1-897522-61-5

The Weekender Effect

Hyperdevelopment in Mountain Towns

Robert William Sandford

ISBN 978-1-897522-10-3

RMB saved the following resources by printing the pages of this book on chlorine-free paper made with 100% post-consumer waste:

Trees · 8, fully grown

Water · 3,519 gallons

Energy · 4 million BTUs

Solid Waste · 223 pounds

Greenhouse Gases · 781 pounds

Calculations based on research by Environmental Defense and the Paper Task Force. Manufactured at Friesens Corporation.